Lecture Notes in

MW00760070

Volume 577

Series Editors

The book series *Lecture Notes in Electrical Engineering* (LNEE) publishes the latest developments in Electrical Engineering - quickly, informally and in high quality. While original research reported in proceedings and monographs has traditionally formed the core of LNEE, we also encourage authors to submit books devoted to supporting student education and professional training in the various fields and applications areas of electrical engineering. The series cover classical and emerging topics concerning:

- Communication Engineering, Information Theory and Networks
- Electronics Engineering and Microelectronics
- Signal, Image and Speech Processing
- Wireless and Mobile Communication
- Circuits and Systems
- Energy Systems, Power Electronics and Electrical Machines
- Electro-optical Engineering
- Instrumentation Engineering
- Avionics Engineering
- Control Systems
- Internet-of-Things and Cybersecurity
- Biomedical Devices, MEMS and NEMS

For general information about this book series, comments or suggestions, please contact leontina. dicecco@springer.com.

To submit a proposal or request further information, please contact the Publishing Editor in your country:

China

Jasmine Dou, Associate Editor (jasmine.dou@springer.com)

India

Swati Meherishi, Executive Editor (swati.meherishi@springer.com)
Aninda Bose, Senior Editor (aninda.bose@springer.com)

Japan

Takeyuki Yonezawa, Editorial Director (takeyuki.yonezawa@springer.com)

South Korea

Smith (Ahram) Chae, Editor (smith.chae@springer.com)

Southeast Asia

Ramesh Nath Premnath, Editor (ramesh.premnath@springer.com)

USA, Canada:

Michael Luby, Senior Editor (michael.luby@springer.com)

All other Countries:

Leontina Di Cecco, Senior Editor (leontina.dicecco@springer.com)
Christoph Baumann, Executive Editor (christoph.baumann@springer.com)

** Indexing: The books of this series are submitted to ISI Proceedings, EI-Compendex, SCOPUS, MetaPress, Web of Science and Springerlink **

More information about this series at http://www.springer.com/series/7818

Ashutosh Kumar Singh · Masahiro Fujita ·
Anand Mohan

Editors

Design and Testing
of Reversible Logic

 Springer

Editors
Ashutosh Kumar Singh
Department of Computer Applications
National Institute of Technology
Kurukshetra, India

Masahiro Fujita
VLSI Design and Education Center
University of Tokyo
Tokyo, Japan

Anand Mohan
Department of Electronics Engineering
Indian Institute of Technology
Varanasi, Uttar Pradesh, India

ISSN 1876-1100   ISSN 1876-1119   (electronic)
Lecture Notes in Electrical Engineering
ISBN 978-981-13-8823-1   ISBN 978-981-13-8821-7   (eBook)
https://doi.org/10.1007/978-981-13-8821-7

This Springer imprint is published by the registered company Springer Nature Singapore Pte Ltd.
The registered company address is: 152 Beach Road, #21-01/04 Gateway East, Singapore 189721, Singapore

Dedicated to
Anushka, Aakash, Akankshya, and Parents.
Ashutosh Kumar Singh

My family, Yuko, Akito, and Kento Fujita.
Masahiro Fujita

My wife: Sudha Mohan, son: Ashish Mohan,
daughter: Amrita Mohan, and Late Parents.
Anand Mohan

Preface

As CMOS scaling is likely to reach its technological barrier in the near future, novel design paradigms are being proposed to keep in pace with the ever-growing need for computational power and speed. Reversible logic circuits (RLCs) provide an alternative and attractive solution to enter into a new era of computation which assists in realizing the imaginations of building ultra-low-power, high-speed, and compact devices and systems. Since all the quantum operations are inherently reversible, these circuits enable the quantum phenomenon of physics in realizing logic circuits, where the measurements can be performed with a higher degree of precision. These circuits are theoretically proven for providing nearly energy-free computation by preventing information loss during operations. The full scope and potential of this technology has probably not yet been imagined, nor will it be until actual hardware is available for reversible computation. A number of physical systems, spanning much of modern physics, are being developed for reversible and quantum computation. Among others, the testing of these circuits has also been a major concern to validate their functionality.

The book provides efficient design, synthesis, and test methodologies for the implementation of reversible logic circuits. The author's contributions in this book are addressed through 14 chapters which meet several challenges stimulated during each work cycle. Also, it has covered several optimal logic designs for quantum dot cellular automata-based circuit. The chapters are organized into four parts, viz. (i) fundamentals of reversible logic, (ii) design and synthesis, (iii) test methodologies, and (iv) quantum cellular automata.

The editors would like to extend their sincere thanks to Dr. Satish Kumar, Director, National Institute of Technology Kurukshetra, India, and Prof. P. K. Jain, Director, IIT (BHU), Varanasi, India, as well as Prof. V. N. Mishra, Head, Department of Electronics Engineering, IIT (BHU), Varanasi, India, for their encouragement and facilitating the access to institutional resources. We also thank

Jitendra Kumar, Ishu Gupta, Rishab Gupta, Deepika Saxena, and Navreet Kaur of Department of Computer Applications, National Institute of Technology, for their constant support in the editorial process.

Kurukshetra, India   Ashutosh Kumar Singh
Tokyo, Japan   Masahiro Fujita
Varanasi, India   Anand Mohan
March 2019

Contents

Editors and Contributors

About the Editors

Ashutosh Kumar Singh is an esteemed researcher and academician in the domain of Electrical and Computer Engineering. Currently, he is working as a Professor and Head, Department of Computer Applications, National Institute of Technology, Kurukshetra, India. He has more than 19 years research, teaching and administrative experience in various University systems of the India, UK, Australia and Malaysia. Dr. Singh obtained his Ph.D. degree in Electronics Engineering from Indian Institute of Technology-BHU, India; Post Doc from Department of Computer Science, University of Bristol, United Kingdom and Charted Engineer from United Kingdom. His research area includes Verification, Synthesis, Design and Testing of Digital Circuits, Predictive Data Analytics, Data Security in Cloud, Web Technology. He has published more than 180 research papers till now in peer-reviewed journals, conferences and news magazines and in these areas. He has also co-authored seven books including "Web Spam Detection Application using Neural Network", "Digital Systems Fundamentals" and "Computer System Organization & Architecture". He has worked as principal investigator/investigator for six sponsored research projects and was a key member on a project from EPSRC (United Kingdom) entitled "Logic Verification and Synthesis in New Framework". Dr. Singh has visited several countries including Australia, United Kingdom, South Korea, China, Thailand, Indonesia, Japan and USA for collaborative research work, invited talks and to present his research work. He had been entitled for 13 awards such as Merit Awards-2003 (Institute of Engineers), Best Poster Presenter-99 in 86th Indian Science Congress held in Chennai, India, Best Paper Presenter of NSC'99 INDIA and Bintulu Development Authority Best Postgraduate Research Paper Award for 2010, 2011, 2012. He has served as an Editorial Board Member of International Journal of Networks and Mobile Technologies, International Journal of Digital Content Technology and its Applications. Also he has shared his experience as a Guest Editor for Pertanika Journal of Science and Technology, Chairman of CUTSE International Conference

2011, Conference Chair of series of International Conference on Smart Computing and Communication (ICSCC), and as editorial board member of UNITAR e-journal. He is involved in reviewing process in different journals and conferences of repute including IEEE transaction of computer, IET, IEEE conference on ITC, ADCOM, etc.

Masahiro Fujita received his Ph.D. in Information Engineering from the University of Tokyo in 1985 on his work on model checking of hardware designs by using logic programming languages. In 1985, he joined Fujitsu as a researcher and started to work on hardware automatic synthesis as well as formal verification methods and tools, including enhancements of BDD/SAT based techniques. From 1993 to 2000, he was director at Fujitsu Laboratories of America and headed a hardware formal verification group developing a formal verifier for real life designs having more than several million gates. The developed tool has been used in production internally at Fujitsu and externally as well. Since March 2000, he has been a professor at VLSI Design and Education Center of the University of Tokyo. He has done innovative work in the areas of hardware verification, synthesis, testing, and software verification-mostly targeting embedded software and web-based programs. He has been involved in a Japanese governmental research project for dependable system designs and has developed a formal verifier for C programs that could be used for both hardware and embedded software designs. The tool is now under evaluation jointly with industry under governmental support. He has authored and co-authored 10 books, and has more than 200 publications. He has been involved as program and steering committee member in many prestigious conferences on CAD, VLSI designs, software engineering, and more. His current research interests include synthesis and verification in SoC (System on Chip), hardware/software co-designs targeting embedded systems, digital/analog co-designs, and formal analysis, verification, and synthesis of web-based programs and embedded programs.

Anand Mohan former Director of National Institute of Technology (NIT), Kurukshetra, Haryana has 41 years of rich experience in teaching, research, industrial R & D. He is currently working as Professor (HAG) in the Department of Electronics Engineering, IIT(BHU), Varanasi. He has made notable research contributions in the areas of robust watermarking, telemedicine, and fault tolerant digital system design. Eleven students have been awarded Ph.D. degree under his supervision. Prof. Mohan has successfully completed eight sponsored projects funded by MHRD, AICTE and Ministry of Communication & Information Technology, Govt. of India, New Delhi. He has published 145 research papers in reputed international/national journals and conference proceedings. Prof. Mohan has provided national level leadership to the defence-related R & D activities as Chairman of Armament Sensors & Electronics (ASE) Panel under ARMREB, DRDO, Ministry of Defence, Govt. of India. He has been associated with several important academic and research advisory committees as Chairman of Ad-hoc committees of Software Technology Parks of India (STPI), DRDO laboratories,

NAAC, and AICTE and Member of DST, UGC, and CSIR committees. He obtained UG, PG and Ph.D. degrees in Electronics Engineering from Banaras Hindu University. He is recipient of 'Life Time Achievement Award' conferred by Kamla Nehru Institute of Technology (KNIT), Sultanpur (2016). He is Fellow of Institution of Electronics and Telecommunication Engineers (IETE), Fellow of Institution of Engineers (I), Member, IEEE, USA life member of Project Management Associates (PMA), New Delhi, and Life member of Indian Society of Technical Education (ISTE), New Delhi.

Contributors

C. Bandyopadhyay Department of Information Technology, Indian Institute of Engineering Science and Technology, Shibpur, India

Debajyoty Banik Indian Institute of Technology Patna, Patna, India

A. Bhattacharjee Department of Information Technology, Indian Institute of Engineering Science and Technology, Shibpur, India

S. C. Chua Intel Corporation, Kuala Lumpur, Malaysia

Rolf Drechsler Institute of Computer Science, University of Bremen & Cyber-Physical Systems, DFKI GmbH, Bremen, Germany

M. Fujita University of Tokyo, Tokyo, Japan

I. Gassoumi Laboratory of Electronics and Microelectronics, University of Monastir, Monastir, Tunisia

H. M. Gaur Department of Electronics & Communication Engineering, ABES Institute of Technology, Ghaziabad (Delhi NCR), India

C. Kalyana Sundaram Mepco Schlenk Engineering College, Sivakasi, India

A. Kamaraj Mepco Schlenk Engineering College, Sivakasi, India

P. Marichamy PSR Engineering College, Sivakasi, India

A. Mohan Department of Electronics Engineering, IIT BHU, Varanasi, India

B. Mondal Department of Information Technology, Indian Institute of Engineering Science and Technology, Shibpur, India

B. Ouni Higher Institute of Technologies of Sousse, University of Sousse, Sousse, Tunisia

S. Parekh Department of Engineering and Architecture, ID-Lab, University of Ghent, Ghent, Belgium

D. K. Pradhan Department of Computer Science, University of Bristol, Bristol, UK

H. Rahaman Department of Information Technology, Indian Institute of Engineering Science and Technology, Shibpur, India

Hafizur Rahaman Indian Institute of Engineering Science and Technology, Shibpur, India

D. Roy Department of Computer Science, University of Central Florida, Orlando, FL, USA

T. N. Sasamal Department of Electronics and Communication Engineering, NIT Kurukshetra, Kurukshetra, India

S. Selva Nidhyananthan Mepco Schlenk Engineering College, Sivakasi, India

J. Senthil Kumar Mepco Schlenk Engineering College, Sivakasi, India

A. K. Singh Department of Computer Applications, NIT Kurukshetra, Kurukshetra, India

L. Touil Laboratory of Electronics and Microelectronics, University of Monastir, Monastir, Tunisia

Robert Wille Institute for Integrated Circuits, Johannes Kepler University Linz, Linz, Austria

Part I
Fundamental Concepts

Reversible Logic: An Introduction

H. M. Gaur, T. N. Sasamal, A. K. Singh, A. Mohan and D. K. Pradhan

Abstract Reversible logic is one of the alternatives to meet the requirement of power, speed and size in EDA (Electronic Design Automation) industry because these circuits are theoretically proven for providing nearly energy free computation by preventing the loss of information during operations. This chapter describes about theory of reversible logic, basic gates, cost matrices used for synthesis and testing of these circuits and connection of reversible logic with quantum computation.

1 Introduction

Energy loss is a significant constraint in digital circuit design. It is involved in each phase of design cycle starting from the logic level to the technological level of development. Since the evolution of electronic devices, starting from the centimeter scale vacuum tubes to the present nanometer scale CMOS: the minimization of power dissipation, lowering the size and enhancing the speed are major challenges in the pursuit of cutting-edge technology. Higher degree of on-chip integrated circuits and type of fabrication processes have dramatically reduced the energy levels over the

H. M. Gaur (✉)
Department of Electronics & Communication Engineering, ABES Institute of Technology, Ghaziabad (Delhi NCR) 201009, India
e-mail: leoharimohan84@gmail.com

T. N. Sasamal
Department of Electronics and Communication Engineering, NIT Kurukshetra, Kurukshetra, India
e-mail: sasamal.trailokyanath@gmail.com

A. K. Singh
Department of Computer Applications, NIT Kurukshetra, Kurukshetra, India
e-mail: ashutosh@nitkkr.ac.in

A. Mohan
Department of Electronics Engineering, IIT-BHU, Varanasi, India
e-mail: profanandmohan@gmail.com

D. K. Pradhan
Department of Computer Science, University of Bristol, Bristol, UK
e-mail: pradhan@compsci.bristol.ac.uk

© Springer Nature Singapore Pte Ltd. 2020
A. K. Singh et al. (eds.), *Design and Testing of Reversible Logic*, Lecture Notes in Electrical Engineering 577, https://doi.org/10.1007/978-981-13-8821-7_1

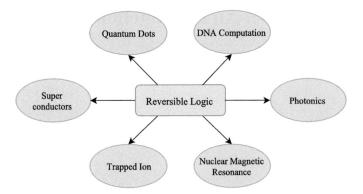

Fig. 1 Computing technologies

last decades. Today, we have achieved in reducing the size to some nanometers scales providing clock speed more than 3 GHz. But if we further minimize the size, we have to compensate with speed and power dissipation. This statement limits the evolution of Moore's law and it may saturate by 2020 [46]. Another factor of power dissipation is the loss of information as highlighted by Landauer [45]. According to this, every bit loss in logic circuits is involved in a loss of $k_B T \ln 2$ Joules of heat in the environment, where k_B is Boltzmann's constant and T is the absolute temperature at which operation is performed. For instance, the heat dissipation per bit is 2.9×10^{-21} Joules [45]. However, it is very small but cannot be ignored where the rate of the processor frequency is very high. This problem propel the engineers and researchers to develop such a logic that does not involve information loss. It is possible only by the inclusion of reversible gates in logic design process. Quantum computation is also known for saving power on the top of several emerging technologies like trapped ion, magnetic resonance etc. Moreover, all the quantum computation are inherently reversible in nature [59].

Reversible logic is one of the promising techniques to reduce the power requirements, as these circuits are theoretically claimed for producing nearly energy free computation systems by preserving the loss of information [6]. These circuits have the capability of producing ultra high speed and compact electronic devices [59]. However, the logic can be applied to traditional logic circuits [93], but its applications to quantum computation have been proven for achieving excellence in terms of power consumption, speed and size. The identification and implementation of reversible quantum circuits have been achieved using several probabilistic methods and ideas [5, 15, 59, 70, 91]. Figure 1 shows some dominating technologies where the researchers are currently exploring the possibilities for employing this logic at physical foregrounds [13, 43, 64, 73, 81].

The construction of reversible logic circuits design and synthesis techniques [14, 22, 26, 31, 48, 55, 65, 66, 78, 79, 92] are based on fundamental Toffoli and Fredkin gates [18, 19, 86]. These gates can be further extended to nth order gates, known as Multiple Control Toffoli (MCT) and Multiple Controlled Fredkin (MCF) gate libraries. Numerous other reversible gates have also been proposed in the literature [7, 32, 33, 39–41, 52, 63, 71, 74, 75, 77, 82, 84], but the primary components of these gates are MCF and MCT [28]. Moreover, the final quantum decomposition of the reversible circuits are based on them. The efficiency of the designs are governed by several performance metrics defining their operating cost. These metrics are wires, gates, quantum stages and garbage [50].

2 The Logic

Currently, all digital logic circuits are physically irreversible because they comprise of irreversible gates. The energy provided by the source is eventually converted into heat with every bit loss. For example, there is a loss of a bit per clock in a two input AND gate as shown in Fig. 2a. Moreover, irreversible logic does not traverse the state sequences in the reverse direction to achieve the initial state after the completion of logical functions. In the I/O of AND gate shown in Table 1, the output permutation is 0 for three input permutations 0, 1 and 2. The output cannot be unique because there are only two logic states (0 and 1) where at least two output will be same, the reverse computation cannot be achieved. The statement cannot be true number of inputs greater than the number of permutations. Hence, there should be same number of inputs and outputs. For two inputs there will be two outputs as shown in Fig. 2b. But assuming two output will not provide the solution. Consider the example circuit of 2×2 AND gate shown in Fig. 2c whose I/O is listed in Table 2. Here, again the output permutation is 0 for two input permutations 0 and 1. There is a need to develop a logic to settle this situation, where we can conserve output bits.

In this situation, information lossless computing presents an option, where a logical operation does not yield loss of information called reversible operation. There

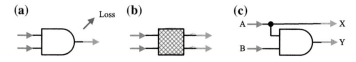

Fig. 2 Irreversible computation

Table 1 I/O of AND gate

Input		Output	
A B	Permutation	Y = A·B	Permutation
0 0	0	0	0
0 1	1	0	0
1 0	2	0	0
1 1	3	1	1

Table 2 I/O of 2 × 2 AND gate

Input		Output		
A B	Permutation	X	Y = A·B	Permutation
0 0	0	0	0	0
0 1	1	0	0	0
1 0	2	1	0	2
1 1	3	1	1	3

Fig. 3 Reversible computation

are two approaches to attain reversibility; they are logical and physical reversibility. The first one corresponds to the bijective relation between inputs and outputs, so inputs can be inferred from the outputs [45]. The later means that there must be some conditions for computation in reverse order [6]. Logical reversibility can be achieved by following two different criteria. First, when intermediate information's are retained during computation from input to output. Here, the reverse computation can be obtained in backward direction, i.e., from output to input by considering the retained information. Second, when logical reversible gates are used for computation without storing intermediate results. However, logical reversibility in turn implies physical reversibility which bards the dissipative effects in computation process. For a computation to be physically reversible, it must be logically reversible. The unsuspecting erasure of a bit of information must always incur a cost of $k_B T \ln 2$ in thermodynamic entropy in irreversible computation. Hence, reversible logic entails the reduction of increment in physical entropy by saving the input information and prevent the information loss in the form of heat to the environment.

Consider the example circuit of 2 × 2 XOR gate shown in Fig. 3a. Its I/O listed Table 3 shows the same results as required for a reversible 2 × 2 function where there

Table 3 I/O of 2 × 2 XOR gate

Input		Output		
A B	Permutation	X = A	Y = A⊕B	Permutation
0 0	0	0	0	0
0 1	1	0	1	1
1 0	2	1	1	3
1 1	3	1	0	2

is a unique correlation between input and output permutations called as bijectivity or logical reversibility. Moreover, it can also be observed from Table 3 that, if the output is again used as an input, which results in original input permutation, called as physical reversibility. For instance, for input permutation 2 the output is 3 and if we use 3 as an input permutation, the output is the original input permutation i.e. 2.

The mathematical equivalence circuit of 2×2 XOR gate shown in Fig. 3b. Its corresponding reversible/quantum representation is shown in Fig. 3c, which is called as reversible Controlled NOT (CNOT) gate. It is having a control input k and target output T. The output function $f(k, T)$ is controlled by the input k. For instance, if $k = 0$, $f(k, T) = T$ and if $k = 1$, $f(k, T) = \overline{T}$, i.e., controlled NOT operation. The concept was very theoretical in the beginning. Many mathematical equations were used to prove its feasibility and strengths. Since the past two decades, this topic drew more attention and improved in design, synthesis and testing. Mainly, there are two categories of current research in this area, physical implementation and designing where a number of research groups in leading research labs around the globe are working.

2.1 Reversible Function

A Boolean function produces p outputs $(y_1, y_2, \ldots, y_p)$ with respect to n inputs $x_1, x_2, \ldots, x_n$, where the output y is the function of inputs $y_p = f(x_1, x_2, \ldots, x_n)$. A Boolean logic function with n variables is *reversible* if it generates unique output for each input permutations [86]. The necessary conditions for a Boolean function with n variables to be reversible can be stated as:

- The number of inputs and outputs should be equal.
- There should be bijective mapping between input and output.

2.2 Reversible Gates

A reversible gate recognizes a reversible function. If a $n \times n$ input output gate produces distinct output for its distinct input functions, it is called as $n \times n$ reversible gate. The basic building blocks for designing any functional circuits are the logic gates and synthesis schemes based on them. Alike irreversible AND, OR and NOT logic gates, there are two fundamental Controlled NOT and Controlled SWAP reversible gates. These gates are commonly known as Toffoli and Fredkin gates, which are further extended into $n \times n$ gates which form Multiple Control Toffoli (MCT) and Multiple Controlled Fredkin (MCF) gates libraries.

Multiple Controlled Toffoli

An MCT gate has m control inputs $(k_1, k_2, \ldots, k_m)$ and one target input T to form $(m + 1) \times (m + 1)$ reversible Boolean function. The control input directly mapped

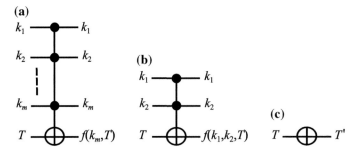

Fig. 4 Schematic representation of MCT gates

to their respective outputs and the function $f(k_m, T)$ is given by Eq. 1. The illustration is also provided in Fig. 4a. A 3×3 gate shown in Fig. 4b is called a Toffoli gate. The least member of this family shown in Fig. 4c is known as a NOT gate, whose output is inversion of the input applied to it.

$$f(k_m, T) = (k_1 \cdot k_2 \cdot \ldots k_m) \oplus T \tag{1}$$

Multiple Controlled Fredkin

An MCF gate has m control inputs $(k_1, k_2, \ldots, k_m)$ and two target inputs T_1 and T_2 to form a $(m + 2) \times (m + 2)$ reversible function as depicted in Fig. 5a. The gate passes all the control inputs directly to respective outputs and the target outputs $f_1(k_m, T_1, T_2)$ and $f_2(k_m, T_1, T_2)$ are given by Eqs. 2 and 3 respectively. Here, $k_{PR} = k_1 \cdot k_2 \cdot \ldots \cdot k_m$. A 3×3 gate shown in Fig. 5b is called a Fredkin gate, whose respective outputs are given by $f_1(k, T_1, T_2) = \overline{k}T_1 + kT_2$ and $f_2(k, T_1, T_2) = kT_1 + \overline{k}T_2$. The least member of this family is shown in Fig. 5c is known as a SWAP gate which interchange its applied target inputs at the outputs.

$$f_1(k_m, T_1, T_2) = \overline{k_{PR}} \cdot T_1 + k_{PR} \cdot T_2 \tag{2}$$

$$f_2(k_m, T_1, T_2) = k_{PR} \cdot T_1 + \overline{k_{PR}} \cdot T_2 \tag{3}$$

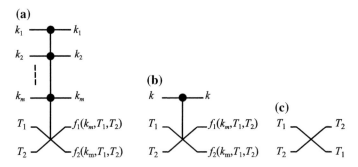

Fig. 5 Schematic representation of MCF gates

Table 4 Input-output of 3×3 gates

Gates	Output		
	P	Q	R
Peres	A	A⊕B	AB⊕C
R	A⊕B	A	C'⊕AB
TR	A	A⊕B	AB'⊕C
URG	A	A⊕B	AB⊕C
PPRG	A⊕B	AC+B'C'	A'C+B'C'

Table 5 Input-output of 4×4 gates

Gates	Output			
	P	Q	R	S
R1	A⊕C	B⊕C⊕AB⊕BC	A⊕B⊕C	D⊕C⊕AB⊕BC
TSG	A	A'B'⊕B'	(A'C'⊕B')⊕D	(A'C'⊕B')D⊕(AB⊕C)
OTG	A	A⊕B	A⊕B⊕D	(A⊕B)D⊕(AB⊕C)
PAOG	A	A⊕B	AB⊕C	((A⊕B)⊕D)⊕(AB⊕C)
SMS	A⊕C⊕D	D⊕BC	C	D⊕B⊕C⊕BC

Others

There are many n I/O gates has been projected in the last two decades other than fundamental MCT and MCF gates, [7, 32, 33, 39–41, 52, 53, 63, 71, 74, 75, 77, 82, 84], which are projected for special functions efficiently. The purposes includes universality, addition, parity preservation etc. The schematics of some popular 3×3 and 4×4 gates are shown in Table 4, where the inputs {A, B, C} and outputs {P, Q, R} are revealed. The schematics of some popular 4×4 gates are shown in Table 5, where the inputs {A, B, C, D} and respective outputs {P, Q, R, S} can be seen. The gates Peres, R and URG are known for their universality. TR, TSG and PAOG gates are projected for addition purposes. The gates SMS, R1, OTG and PPRG are having parity preservation and generation capabilities which can be used for testability purpose. The properties of nearly all commonly used 3×3 and 4×4 gates are summarized and analyzed at two significant experimentation levels in [28].

2.3 Fundamental Properties of Reversible Gates and Circuits

Depending on the functionality of reversible gates, they can be categorized into two different classes:

- *Conservative*: The gates which retain the number of logic values from the input to output are known as conservative gates. In other words, the number of 1s and 0s are same at both ends of the circuit.

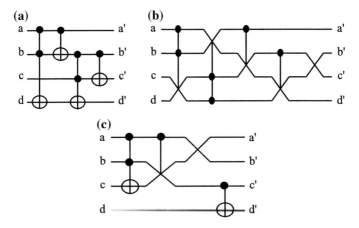

Fig. 6 Reversible circuits

- *Parity preserving*: The gates in which the sum of logic $1s$ in their inputs and outputs are even are called parity preserving gates. Scientifically, the exor of all inputs and outputs result a null value. If these properties are closely analyzed it can be concluded that the conservative gates are parity preserving but vice-versa is not always correct.

The construction of all reversible logic circuits design and synthesis techniques are based on fundamental MCT and MCF gates. Since the operations in reversible logic circuits are linear, i.e., all the outputs of one gate stage varies the operations of next gate stages. The input signal is once propagated from the input, will not be taken back and cannot be taken as inputs to multiple gates and there is no loss of logic bits which maintains the information entropy. The synthesis of reversible circuits is restricted to 'FANOUT' and 'FEEDBACK'. In reversible circuits, these factors are not allowed [6, 59]. It results in the only condition to form a reversible circuit by means of cascading reversible gates, Fig. 6a, b, c illustrated the three reversible circuit networks by using MCT, MCF and MCTF gates respectively.

2.4 Cost Metrics

In designing reversible circuits, certain cost measures has been considered to evaluate the efficacy of proposed methodologies. These measures are can be considered as operating cost of the circuits [23, 24, 50]. A brief explanation and respective illustration using a reversible full adder (*rd*32 benchmark circuit [16] shown in Fig. 7) of these metrics is given as follows:

Fig. 7 Reversible full adder

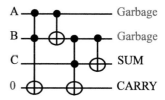

Fig. 8 Schematic of
reversible full adder

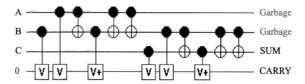

Gate Cost:

The number of gates required to construct a circuit refer its gate count. It is a direct measure to calculate the cost of a circuit, which is commonly called as gate cost (GC). The number of gates used to construct a full adder shown in Fig. 7 are four, hence its gate cost is equal to 4.

Quantum Cost:

A complete reversible circuit can also be realized in corresponding quantum realization using elementary quantum gates (1×1 NOT, 1×1 CNOT, Control V and Control V+). The sum of these elementary quantum gates are termed as quantum cost (QC) of the circuit. The full adder (Fig. 7) can be decomposed using twelve quantum gates as shown in Fig. 8, hence its quantum cost is 12.

Ancilla Input:

There are the requirement of additional constant inputs to convert an irreversible into reversible circuit. These additional inputs are referred as ancilla inputs (AI). The last input is taken as constant 0 to construct SUM and CARRY functions as shown in green colour in Fig. 7, hence number of ancilla input is 1.

Input:

The number of inputs directly impact on the qubits in a reversible circuit and increases the size of the circuit. These number of input or wires (n) including ancilla input can also be taken to evaluate the performance of any design methodology. There are four wires needed to implement a full adder circuit shown in Fig. 7, hence $n = 4$.

Garbage Output:

To maintain the bijectivity property in the realization of reversible circuits, some outputs are left unused. These outputs are referred as garbage outputs (GO). It can be seen that, two output are left unused to realize a full adder circuit depicted in Fig. 7, written as garbage in red colour. Hence GO for this circuit is 2.

The discussed parameters show a direct connection with area/size and complexity of a circuit. The more values of these parameters raise the size and complexity and enhance the power dissipation and physical cost of circuits and devices. Hence the minimization of these measures should be the major subject all through the development of design and test strategy for reversible logic circuits. The quantum cost has a proportionate association with delay, while ancilla inputs are the extra source of input power and garbage depicts the loss of power. Moreover, there are some technology-dependent measures are also exist like input to output delay which can be used to compute the speed.

3   A Door Step Towards Quantum Computation

A traditional computer circuits comprised of wires and logic gates, where the wires transfer the information which is further manipulated by logic gates to perform some operations. In quantum computers, the information is switched with the help of change in quantum states. These states are defined using linear combinations, called as superposition for a single *qu*bit system as given in Eq. 4. Where, α & β are the two complex numbers and $|0\rangle$, $|1\rangle$ shows the Dirac notations of the two states.

$$|\psi\rangle = \alpha|0\rangle + \beta|1\rangle \tag{4}$$

Out of the infinite quantum information space (Hilbert Space) in the superposition, there are two quantum states $|0\rangle$ and $1\rangle$ that can be easily recognizable for performing any operation. These states refers the ground state and excited state of electrons in an atom respectively, as demonstrated in Fig. 9. The *qu*bit is once propagated from the input will not be taken back and cannot be taken as inputs to multiple stages in a quantum system.

A quantum circuit contains elementary quantum gates to hold and manipulate the quantum information. The smallest (and trivial) member of this family is elementary NOT gate whose functionality is as usual. Rather than a truth table, these gates are represented in form of a matrix, which pursue the linearity of quantum gates. Controlled NOT, Z, H, SWAP are the elementary quantum gates that can be derived to multiple *qu*bit gates. Similar to reversible gates a quantum controlled NOT gate used to invert a state and SWAP gate is used to interchange any two states. The change of

Fig. 9 *qu*bit representation
as two electronic levels

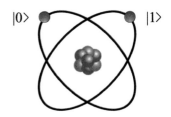

$|0\rangle$ $|1\rangle$

states in quantum NOT gate can be given by $\alpha|0\rangle + \beta|1\rangle \rightarrow \alpha|1\rangle + \beta|0\rangle$. The swap and copy operations are process for minimum of two *qu*bit system. The archetypal multi-*qu*bit quantum gates are controlled NOT gates, similar as that of reversible MCT and MCF gates and all the quantum operations are inherently reversible in nature.

Hence, regardless of the physical implementation, the research in design and synthesis of reversible logic circuits will be the foundations of quantum computers. The implementation of reversible logic will be formulated in conjunction with quantum technology. The full scope of potential technologies has not been imagined yet nor it will be, until definite quantum information hardware is obtainable for future generations of computation. The reversible logic designs and synthesis methods are independent of the implementation technique. Once the implementation becomes more stable for fabrication, the design can be set into a real hardware device.

4   Summary of the Chapter

However, physical implementation is limitedly experimented, a healthy research has been accomplished in the area of reversible logic circuits. Following are the key points that are discussed in this chapter:

- Description of the theory of reversible logic
- Basis definitions and notations
- Explanations of reversible gates and their properties
- Explains all the cost metrics used for analysis
- Efficacy of Toffoli and Fredkin gates for reversible and quantum circuit implementations ihas been emphasized
- Connections of reversible logic with quantum computation

References

1. Abdessaied N, Amy M, Soeken M, Drechsler R (2016) Technology mapping of reversible circuits to Clifford+T quantum circuits. In: 2016 IEEE 46th international symposium on multiple-valued logic (ISMVL), pp 150–155. https://doi.org/10.1109/ISMVL.2016.33
2. Amy M, Maslov D, Mosca M, Roetteler M (2013) A meet-in-the-middle algorithm for fast synthesis of depth-optimal quantum circuits. IEEE Trans Comput-Aided Des Integr Circuits Syst 32(6):818–830. https://doi.org/10.1109/TCAD.2013.2244643
3. Avizienis A (1978) Fault-tolerance: the survival attribute of digital systems. Proc IEEE 66(10):1109–1125. https://doi.org/10.1109/PROC.1978.11107
4. Babu HMH, Mia MS, Biswas AK (2017) Efficient techniques for fault detection and correction of reversible circuits. J Electron Test 1–15. https://doi.org/10.1007/s10836-017-5679-4
5. Barenco A, Bennett CH, Cleve R, DiVincenzo DP, Margolus N, Shor P, Sleator T, Smolin JA, Weinfurter H (1995) Elementary gates for quantum computation, vol 52

6. Bennett CH (1973) Logical reversibility of computation. IBM J Res Dev 17(6):525–532. https://doi.org/10.1147/rd.176.0525
7. Biswas AK, Hasan MM, Chowdhury AR, Babu HMH (2008) Efficient approaches for designing reversible binary coded decimal adders. Microelectron J 39(12):1693–1703. https://doi.org/10.1016/j.mejo.2008.04.003, http://www.sciencedirect.com/science/article/pii/S0026269208001791
8. Boykin PO, Roychowdhury VP (2005) Reversible fault-tolerant logic. In: 2005 International conference on dependable systems and networks (DSN'05), pp 444–453. https://doi.org/10.1109/DSN.2005.83
9. Bubna M, Goyal N, Sengupta I (2007) A DFT methodology for detecting bridging faults in reversible logic circuits. In: TENCON 2007–2007 IEEE region 10 conference, pp 1–4. https://doi.org/10.1109/TENCON.2007.4428915
10. Burchard J, Erb D, Singh AD, Reddy SM, Becker B (2017) Fast and waveform-accurate hazard-aware SAT-based TSOF ATPG. In: Design, automation test in Europe conference exhibition (DATE) 2017, pp 422–427. https://doi.org/10.23919/DATE.2017.7927027
11. Chakraborty A (2005) Synthesis of reversible circuits for testing with universal test set and C-testability of reversible iterative logic arrays. In: 18th International conference on VLSI design held jointly with 4th International conference on embedded systems design, pp 249–254. https://doi.org/10.1109/ICVD.2005.158
12. Chakraborty A (2010) Testing of bridging faults in AND-EXOR based reversible logic circuits. CoRR arXiv:abs/1009.5098, http://arxiv.org/abs/1009.5098
13. Chen JL, Zhang XY, Wang LL, Wei XY, Zhao WQ (2008) Extended Toffoli gate implementation with photons. In: 2008 9th international conference on solid-state and integrated-circuit technology, pp 575–578. https://doi.org/10.1109/ICSICT.2008.4734595
14. Cheng CS, Singh AK, Gopal L (2015) Efficient three variables reversible logic synthesis using mixed-polarity Toffoli gate. In: Proceedings of the 4th international conference on eco-friendly computing and communication systems; Procedia Comput Sci 70:362 – 368. https://doi.org/10.1016/j.procs.2015.10.035, http://www.sciencedirect.com/science/article/pii/S1877050915031993
15. Chiribella G, D'Ariano GM, Roetteler M (2013) Identification of a reversible quantum gate: assessing the resources. New J Phys 15(10):103019. http://stacks.iop.org/1367-2630/15/i=10/a=103019
16. D Maslov, G and Dueck, N Scott (2004) Reversible logic synthesis benchmarks page. http://webhome.cs.uvic.ca/~dmaslov/. Accessed: 23 Sept 2017
17. Farazmand N, Zamani M, Tahoori MB (2010) Online fault testing of reversible logic using dual rail coding. In: 2010 IEEE 16th international on-line testing symposium, pp 204–205. https://doi.org/10.1109/IOLTS.2010.5560205
18. Feynman RP (1986) Quantum mechanical computers. Found Phys 16(6):507–531. https://doi.org/10.1007/BF01886518
19. Fredkin E, Toffoli T (1982) Conservative logic. Int J Theor Phys 21(3):219–253. https://doi.org/10.1007/BF01857727
20. Fujita M (2015a) Automatic identification of assertions and invariants with small numbers of test vectors. In: 2015 33rd IEEE international conference on computer design (ICCD), pp 463–466. https://doi.org/10.1109/ICCD.2015.7357149
21. Fujita M (2015b) Detection of test patterns with unreachable states through efficient inductive-invariant identification. In: 2015 IEEE 24th Asian test symposium (ATS), pp 31–36. https://doi.org/10.1109/ATS.2015.13
22. Gaur HM, Singh AK (2016) Design of reversible circuits with high testability. Electron Lett 52(13):1102–1104. https://doi.org/10.1049/el.2016.0161
23. Gaur HM, Singh AK, Ghanekar U (2015) A review on online testability for reversible logic. Procedia Comput Sci 70:384–391
24. Gaur HM, Singh AK, Ghanekar U (2016a) A comprehensive and comparitive study on online testability in reversible logic. Pertanika J Sci Technol 24(2):245–271

25. Gaur HM, Singh AK, Ghanekar U (2016b) A new DFT methodology for k-CNOT reversible circuits and its implementation using quantum-dot cellular automata. Optik - Int J Light Electron Opt 127(22):10593–10601. https://doi.org/10.1016/j.ijleo.2016.08.072, http://www.sciencedirect.com/science/article/pii/S003040261630941X
26. Gaur HM, Singh AK, Ghanekar U (2017) Testable design of reversible circuits using parity preserving gates. IEEE Des Test
27. Gaur HM, Singh AK, Ghanekar U (2018a) Design for stuck-at faults testability in MCT based reversible circuits. Def Sci J 68(4):381–387
28. Gaur HM, Singh AK, Ghanekar U (2018b) In-depth comparative analysis of reversible gates for designing logic circuits. Procedia Comput Sci 125:810–817
29. Gaur HM, Singh AK, Ghanekar U (2018c) Offline testing of reversible logic circuits: an analysis. Integration 62:50–67. https://doi.org/10.1016/j.vlsi.2018.01.004, http://www.sciencedirect.com/science/article/pii/S0167926016301341
30. Gaur HM, Singh AK, Ghanekar U (2018d) Reversible circuits with testability using quantum controlled NOT and swap gates. Indian J Pure Appl Phys 56(7):529–532
31. Golubitsky O, Maslov D (2012) A study of optimal 4-bit reversible Toffoli circuits and their synthesis. Indian J Pure Appl Phys 61(9):1341–1353. https://doi.org/10.1109/TC.2011.144
32. Haghparast M, Navi K (2007) A novel reversible full adder circuit for nanotechnology based systems. J Appl Sci 7(24):3995–4000
33. Haghparast M, Navi K (2008) A new reversible design of bcd adder. Am J Appl Sci 5:282–288. https://doi.org/10.3844/ajassp.2008.282.288
34. Hasan M, Islam AKMT, Chowdhury AR (2009) Design and analysis of online testability of reversible sequential circuits. In: 2009 12th international conference on computers and information technology, pp 180–185. https://doi.org/10.1109/ICCIT.2009.5407143
35. Hayes JP, Polian I, Becker B (2004) Testing for missing-gate faults in reversible circuits. In: 13th Asian test symposium, pp 100–105. https://doi.org/10.1109/ATS.2004.84
36. Hurst SL (1998) VLSI Testing: Digital and Mixed Analogue/Digital Techniques. The Institution of Electrical Engineers, London
37. Ibrahim M, Chowdhury AR, Babu HMH (2008) Minimization of CTS of k-CNOT circuits for SSF and MSF model. In: 2008 IEEE international symposium on defect and fault tolerance of VLSI systems, pp 290–298. https://doi.org/10.1109/DFT.2008.38
38. iNEMI-Roadmap Executive Summary Highlights (2015) International Electronics Manufacturing Initiative
39. Islam MS (2010) A novel quantum cost efficient reversible full adder gate in nanotechnology. CoRR arxiv.abs/1008.3533
40. Islam MS, Rahman M, Begum Z, Hafiz MZ (2009a) Low cost quantum realization of reversible multiplier circuit. Inf Technol J 8(2):208–213
41. Islam MS, Rahman MM, Begum Z, Hafiz MZ (2009b) Fault tolerant reversible logic synthesis: carry look-ahead and carry-skip adders. In: 2009 international conference on advances in computational tools for engineering applications, pp 396–401. https://doi.org/10.1109/ACTEA.2009.5227871
42. Jha NK, Gupta S (2003) Testing of digital systems. Cambridge University Press, Cambridge
43. Lm K, Vandersypen Steffen M, Breyta G, Yannoni CS, Sherwood MH, Chuang IL (2001) Experimental realization of shor's quantum factoring algorithm using nuclear magnetic resonance. Nature 414:883–887. https://doi.org/10.1038/414883a
44. Kole DK, Rahaman H, Das DK, Bhattacharya BB (2011) Derivation of automatic test set for detection of missing gate faults in reversible circuits. In: 2011 international symposium on electronic system design (ISED), pp 200–205. IEEE
45. Landauer R (1961) Irreversibility and heat generation in the computing process. IBM J Res Dev 5(3):183–191. https://doi.org/10.1147/rd.53.0183
46. Mack C (2015) The multiple lives of Moore's law. IEEE Spectr 52(4):31–31. https://doi.org/10.1109/MSPEC.2015.7065415
47. Mahammad SN, Veezhinathan K (2010) Constructing online testable circuits using reversible logic. IEEE Trans Instrum Meas 59(1):101–109. https://doi.org/10.1109/TIM.2009.2022103

48. Mathew J, Rahaman H, Jose BR, Pradhan DK (2008a) Design of reversible finite field arithmetic circuits with error detection. In: 21st international conference on VLSI design (VLSID 2008), pp 453–459. https://doi.org/10.1109/VLSI.2008.96

49. Mathew J, Singh J, Taleb AA, Pradhan DK (2008b) Fault tolerant reversible finite field arithmetic circuits. In: 2008 14th IEEE international on-line testing symposium, pp 188–189. https://doi.org/10.1109/IOLTS.2008.35

50. Mohammadi M, Eshghi M (2009) On figures of merit in reversible and quantum logic designs. Quantum Inf Process 8(4):297–318. https://doi.org/10.1007/s11128-009-0106-0

51. Mondal B, Kole DK, Das DK, Rahaman H (2014) Generator for test set construction of smgf in reversible circuit by boolean difference method. In: 2014 IEEE 23rd Asian test symposium, pp 68–73. https://doi.org/10.1109/ATS.2014.24

52. Morrison M, Ranganathan N (2011a) Design of a reversible ALU based on novel programmable reversible logic gate structures. In: 2011 IEEE computer society annual symposium on VLSI, pp 126–131. https://doi.org/10.1109/ISVLSI.2011.30

53. Morrison M, Ranganathan N (2011b) Design of a reversible alu based on novel programmable reversible logic gate structures. In: 2011 IEEE computer society annual symposium on VLSI, pp 126–131. https://doi.org/10.1109/ISVLSI.2011.30

54. Nayeem NM, Rice JE (2011a) Online fault detection in reversible logic. In: 2011 IEEE international symposium on defect and fault tolerance in VLSI and nanotechnology systems, pp 426–434. https://doi.org/10.1109/DFT.2011.55

55. Nayeem NM, Rice JE (2011b) A shared-cube approach to ESOP-based synthesis of reversible logic. Facta Univ-Ser: Electron Energ 24(3):385–402

56. Nayeem NM, Rice JE (2011c) A simple approach for designing online testable reversible circuits. In: Proceedings of 2011 IEEE Pacific rim conference on communications, computers and signal processing, pp 85–90. https://doi.org/10.1109/PACRIM.2011.6032872

57. Nayeem NM, Rice JE (2012) A new approach to online testing of tgfsop-based ternary toffoli circuits. In: 2012 IEEE 42nd international symposium on multiple-valued logic, pp 315–321. https://doi.org/10.1109/ISMVL.2012.57

58. Nayeem NM, Rice JE (2013) Online testable approaches in reversible logic. J Electron Test 29(6):763–778. https://doi.org/10.1007/s10836-013-5399-3

59. Nielsen MA, Chuang IL (2011) Quantum Computation and Quantum Information, 10th edn. Cambridge University Press, New York

60. Parhami B (2006) Fault-tolerant reversible circuits. In: 2006 fortieth asilomar conference on signals, systems and computers, pp 1726–1729. https://doi.org/10.1109/ACSSC.2006.355056

61. Patel KN, Hayes JP, Markov IL (2003) Fault testing for reversible circuits. In: Proceedings of 21st VLSI test symposium, pp 410–416. https://doi.org/10.1109/VTEST.2003.1197682

62. Patel KN, Hayes JP, Markov IL (2004) Fault testing for reversible circuits. IEEE Trans Comput-Aided Des Integr Circuits Syst 23(8):1220–1230. https://doi.org/10.1109/TCAD.2004.831576

63. Peres A (1985) Reversible logic and quantum computers. Phys Rev A 32:3266–3276. https://doi.org/10.1103/PhysRevA.32.3266

64. Polian I, Fiehn T, Becker B, Hayes JP (2005) A family of logical fault models for reversible circuits. In: 14th Asian test symposium (ATS'05), pp 422–427. https://doi.org/10.1109/ATS.2005.9

65. Prasad PWC, Singh AK, Beg A, Assi A (2006) Modelling the xor/xnor boolean functions complexity using neural network. In: 2006 13th IEEE international conference on electronics, circuits and systems, pp 1348–1351. https://doi.org/10.1109/ICECS.2006.379732

66. Prasad PWC, Assi A, Beg A (2007) Binary decision diagrams and neural networks. J Supercomput 39(3):301–320. https://doi.org/10.1007/s11227-006-0010-7

67. Rahaman H, Kole DKKDK, Das DK, Bhattacharya BB (2007) Optimum test set for bridging fault detection in reversible circuits. In: 16th Asian test symposium (ATS 2007), pp 125–128. https://doi.org/10.1109/ATS.2007.91

68. Rahaman H, Kole DK, Das DK, Bhattacharya BB (2008) On the detection of missing-gate faults in reversible circuits by a universal test set. In: 21st international conference on VLSI design (VLSID 2008), pp 163–168. https://doi.org/10.1109/VLSI.2008.106

69. Rahaman H, Kole DK, Das DK, Bhattacharya BB (2011) Fault diagnosis in reversible circuits under missing-gate fault model. Comput Electr Eng 37(4):475–485. https://doi.org/10.1016/j.compeleceng.2011.05.005

70. Rentergem YV, Vos AD, Storme L (2005) Implementing an arbitrary reversible logic gate. J Phys A: Math Gen 38(16):3555. http://stacks.iop.org/0305-4470/38/i=16/a=007

71. Roohi A, Zand R, Angizi S, Demara RF (2016) A parity-preserving reversible QCA gate with self-checking cascadable resiliency. IEEE Trans Emerg Top Comput (99):1–1. https://doi.org/10.1109/TETC.2016.2593634

72. Sarkar P, Chakrabarti S (2008) Universal test set for bridging fault detection in reversible circuit. In: 2008 3rd international design and test workshop, pp 51–56. https://doi.org/10.1109/IDT.2008.4802464

73. Sarker A, Babu HMH, Rashid SMM (2015) Design of a DNA-based reversible arithmetic and logic unit. IET Nanobiotechnology 9(4):226–238. https://doi.org/10.1049/iet-nbt.2014.0056

74. Sasamal TN, Mohan A, Singh AK (2016a) Design of parity preserving combinational circuits using reversible gate. In: 2016 2nd international conference on next generation computing technologies (NGCT), pp 631–638. https://doi.org/10.1109/NGCT.2016.7877489

75. Sasamal TN, Singh AK, Mohan A (2016b) Efficient design of reversible alu in quantum-dot cellular automata. Optik - Int J Light Electron Opt 127(15):6172–6182. https://doi.org/10.1016/j.ijleo.2016.04.086, http://www.sciencedirect.com/science/article/pii/S0030402616303618

76. Sen B, Das J, Sikdar BK (2012) A dft methodology targeting online testing of reversible circuit. In: 2012 international conference on devices, circuits and systems (ICDCS), pp 689–693. https://doi.org/10.1109/ICDCSyst.2012.6188661

77. Sen B, Dutta M, Goswami M, Sikdar BK (2014) Modular design of testable reversible ALU by QCA multiplexer with increase in programmability. Microelectron J 45(11):1522–1532. https://doi.org/10.1016/j.mejo.2014.08.012, http://www.sciencedirect.com/science/article/pii/S0026269214002663

78. Shende VV, Prasad AK, Markov IL, Hayes JP (2002) Reversible logic circuit synthesis. In: IEEE/ACM international conference on computer aided design, ICCAD 2002, pp 353–360. https://doi.org/10.1109/ICCAD.2002.1167558

79. Shende VV, Prasad AK, Markov IL, Hayes JP (2003) Synthesis of reversible logic circuits. IEEE Trans Comput-Aided Des Integr Circuits Syst 22(6):710–722. https://doi.org/10.1109/TCAD.2003.811448

80. Tabei K, Yamada T (2009) On generating test sets for reversible circuits. In: 2009 international conference on computer engineering systems, pp 94–99. https://doi.org/10.1109/ICCES.2009.5383305

81. Takeuchi N, Yamanashi Y, Yoshikawa N (2014) Reversible logic gate using adiabatic super-conducting devices. Sci Rep, Nat 4(6354). https://doi.org/10.1038/srep06354

82. Thapliyal H, Ranganathan N (2009) Design of efficient reversible binary subtractors based on a new reversible gate. In: 2009 IEEE computer society annual symposium on VLSI, pp 229–234. https://doi.org/10.1109/ISVLSI.2009.49

83. Thapliyal H, Ranganathan N (2010) Reversible logic based concurrent error detection methodology for emerging nanocircuits. In: 10th IEEE international conference on nanotechnology, pp 217–222. https://doi.org/10.1109/NANO.2010.5697743

84. Thapliyal H, Vinod AP (2006) Transistor realization of reversible tsg gate and reversible adder architectures. In: APCCAS 2006-2006 IEEE Asia pacific conference on circuits and systems, pp 418–421. https://doi.org/10.1109/APCCAS.2006.342478

85. Thapliyal H, Vinod AP (2007) Designing efficient online testable reversible adders with new reversible gate. In: 2007 IEEE international symposium on circuits and systems, pp 1085–1088. https://doi.org/10.1109/ISCAS.2007.378198

86. Toffoli T (1980) Reversible computing. In: de Bakker J, van Leeuwen J (eds) Automata, languages and programming. Springer, Berlin, pp 632–644

87. Vasudevan DP, Lala PK, Parkerson JP (2004) Online testable reversible logic circuit design using NAND blocks. In: Proceedings of 19th IEEE international symposium on defect and fault tolerance in VLSI systems, DFT 2004, pp 324–331. https://doi.org/10.1109/DFTVS.2004.1347856

88. Vasudevan DP, Lala PK, Parkerson JP (2005a) The construction of a fault tolerant reversible gate for quantum computation. In: 5th IEEE conference on nanotechnology, vol 1, pp 112–115. https://doi.org/10.1109/NANO.2005.1500705

89. Vasudevan DP, Lala PK, Parkerson JP (2005b) Fault tolerant quantum computation with new reversible gate. In: Technical proceedings of the 2005 NSTI nanotechnology conference and trade show, vol 3, pp 744 – 747

90. Vasudevan DP, Lala PK, Di J, Parkerson JP (2006) Reversible-logic design with online testability. IEEE Trans Instrum Meas 55(2):406–414. https://doi.org/10.1109/TIM.2006.870319

91. Vos AD, Rentergem YV, Keyser KD (2006) The decomposition of an arbitrary reversible logic circuit. J Phys A: Math Gen 39(18):5015. http://stacks.iop.org/0305-4470/39/i=18/a=017

92. Wille R, Soeken M, Miller DM, Drechsler R (2014) Trading off circuit lines and gate costs in the synthesis of reversible logic. Integr VLSI J 47(2):284–294. https://doi.org/10.1016/j.vlsi. 2013.08.002, http://www.sciencedirect.com/science/article/pii/S0167926013000436

93. Ye Y, Roy K (1996) Energy recovery circuits using reversible and partially reversible logic. IEEE Trans Circuits Syst I: Fundam Theory Appl 43(9):769–778. https://doi.org/10.1109/81. 536746

94. Zamani M, Tahoori MB (2011) Online missing/repeated gate faults detection in reversible circuits. In: 2011 IEEE international symposium on defect and fault tolerance in VLSI and nanotechnology systems, pp 435–442. https://doi.org/10.1109/DFT.2011.56

95. Zamani M, Farazmand N, Tahoori MB (2011) Fault masking and diagnosis in reversible circuits. In: 2011 16th IEEE European test symposium, pp 69–74. https://doi.org/10.1109/ETS.2011. 19

96. Zamani M, Tahoori MB, Chakrabarty K (2012) Ping-pong test: Compact test vector generation for reversible circuits. In: 2012 IEEE 30th VLSI test symposium (VTS), pp 164–169. https:// doi.org/10.1109/VTS.2012.6231097

97. Zhong J, Muzio JC (2006) Analyzing fault models for reversible logic circuits. In: 2006 IEEE international conference on evolutionary computation, pp 2422–2427. https://doi.org/10.1109/ CEC.2006.1688609

Part II
Design & Synthesis

Design of Reversible Hardware BinDCT

I. Gassoumi, L. Touil and B. Ouni

Abstract Recently, reversible logic computation has attracted researchers' attention for implementing low-power digital logic designs. In fact, no information is wasted in this approach, i.e., it performs a bijective function. This chapter introduces a hardware design of reversible BinDCT. It is a new proposal in reversible approach. In this study, we dealt with a variety of sub-modules, which have a better performance in terms of constant inputs (CIs), garbage output (GO), power and quantum cost (QC) as well as the delay than that of existing designs. This work can offer a vital step in the design of reversible designs for in the field of image processing. It could also be present as an essential step in this area since the image processing systems are known to be the biggest energy consumers.

1 Introduction

As we integrate more and more logic elements into lower volumes and pilot them at high frequencies, the system dissipates more and more heat. This creates many issues like portable systems exhaust their batteries, energy costs money, systems overheat, and others. In the classical digital circuits, significant energy dissipation appears, i.e., irreversible devices could losses some information over the treatments. Information waste produces because the outputs signals do not uniquely define its inputs signals. In fact, from Landauer's principle, the energy transferred in the form of heat is at least 0.6931 K.T joules where "T" is the temperature of the environment [1]. In reversible logic computation, there are no erased bits [2]. One of the emerging technologies that can be used for building partially low-power digital systems is reversible logic.

I. Gassoumi (✉) · L. Touil
Laboratory of Electronics and Microelectronics, University of Monastir, Monastir, Tunisia
e-mail: gassoumiismail@gmail.com

L. Touil
e-mail: lamjedtl@yahoo.fr

B. Ouni
Higher Institute of Technologies of Sousse, University of Sousse, Sousse, Tunisia
e-mail: ouni_bouraoui@yahoo.fr

© Springer Nature Singapore Pte Ltd. 2020
A. K. Singh et al. (eds.), *Design and Testing of Reversible Logic*, Lecture Notes in Electrical Engineering 577, https://doi.org/10.1007/978-981-13-8821-7_2

Thus to save power dissipation, systems should be performed from gates reversible in nature. A circuit is reversible if its output values define its input values, i.e., it performs a bijective function since it can return to its initial state [2–5].

Synthesis of reversible designs imposes more design constraints than traditional irreversible synthesis. Reversible circuit is a circuit that is able to return to any previous state in reverse order. In recent years, several efforts have been made toward the design of many arithmetic circuits such as adders/subtractors, BCD adders, multipliers, and other digital circuits [6–17]. On the other hand, for numerical circuits, video processing systems are one of the most energy consumers' applications. Discrete Cosine Transform (DCT) is one of the most important modules of several signal processing systems such as image and audio compression [18]. It consumes large amounts of energy because of the intensive computations required [18, 19]. Therefore, to further reduce the complexity of the DCT module, some approximations have been proposed by researchers to tackle this problem [20]. The BinDCT is fast multiplierless approximations of DCT [21], which is composed of adders and shift registers. In the last few years, several works are devoted to VLSI implementation of the BinDCT algorithm [22–24]. Most of these works are concerned in the classical implementation of this algorithm. However, the semiconductor/VLSI industry faces problems in the domain of device density, short channel effects, and scaling along with power consumption. Consequently, these truths motivate designers to study new solutions to grant low power consumption for image processing systems. Reversible approach is one of the most promising technologies, which can present a fundamental step toward the design of future digital circuits.

Recently, some interesting reversible designs have been proposed for image processing applications [25, 26]. All these above factors motivate us to investigate a new architecture around reversible logic, which can efficiently perform BinDCT operation.

This chapter is organized as follows: In Sect. 2, basic reversible systems are introduced. In Sect. 3, the architecture of BinDCT algorithm is presented. Section 4 describes the implementation of reversible BinDCT module. Results and comparison of the proposed design are reported in Sect. 5.

2 Basic Reversible Logic Gates and Literature Overview

Till now, researchers/designers are devoted to propose reversible gates [27–38]. Each gate has a related quantum cost (QC). The NOT gate is a 1-q-bit gate and it has a QC of zero. The N-bit Controlled-Gate has QC of n−1. The Feynman gate can function as a CNOT. It is extensively utilized to surmount the fan-out issue since it is not permitted in the reversible approach. The QC of Feynman gate is one. The QC of a Double Feynman gate is 2. In addition, TR gate, Peres gate, and Fredkin gate are 3* 3 reversible gates. The QCs of mentioned gates are, respectively, five, four, and four. On the other hand, the delay presents a fundamental metric of a logic design. It

Table 1 Reversible logic gates

Gate	Quantum cost	Delay
Feynman gate	1	$\Delta_{FG} = 1\Delta$
NOT gate	1	$\Delta_{NG} - 1\Delta$
TR gate	5	$\Delta_{TRG} = 4\Delta$
Toffoli gate	4	$\Delta_{TG} = 5\Delta$
Peres gate	4	$\Delta_{PG} = 4\Delta$
Fredkin gate	5	$\Delta_{FRG} = 5\Delta$
BHA gate	4	$\Delta_{BHA} = 4\Delta$

may be determined by the utmost number of gates in the critical path [39]. The total delay of the circuit is Delay = Depth*Δ

Reversible gates (1*1 and 2*2) are generally used to determine the logical depth [39, 40]. The quantum gates (V and V+) can be determined by the subsequent properties:

$V*V = NOT$
$V*V+ = V+ = V = I$
$V+*V+ = Not$

Table 1 shows the QC and the delay of some reversible gates. Figure 1 depicts some reversible gates and its quantum realization. These reversible gates help researchers/designers to design higher complex computing circuits.

The goodness of a reversible circuit is specified by certain criteria as follows:

- Constant inputs (CIs): Some inputs values ("0" and "1") are added in the circuit to perform desired function.
- A complete reversible design may also be realized in corresponding quantum implementation using elementary quantum gates. The sums of these gates are noted as quantum cost (QC) of the system.
- Garbage outputs (GOs): To retain the bijectivity property in the achievement of reversible designs, some outputs are unused. These outputs are known as garbage outputs.
- Gate count (GC): The whole number of reversible gates needed to realize a reversible system corresponds to its gate count.

3 BinDCT Algorithm

As the DCT computation imposes complicated calculation, its hardware implementation consumes further power and area [41]. The BinDCT approximation reduces the complication of the DCT module. Several algorithms for BinDCT have been proposed in the literature. These configurations present a difference in the number

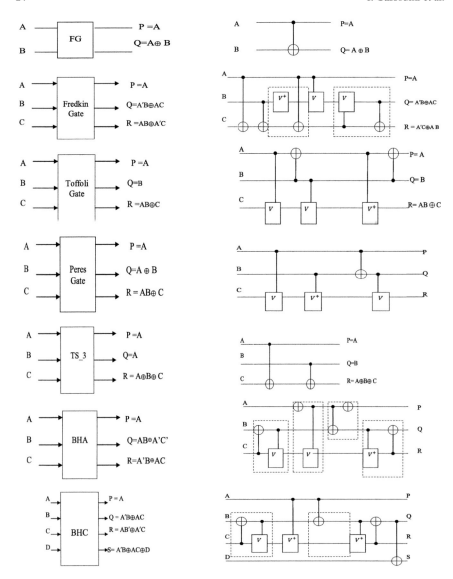

Fig. 1 Reversible gates and its quantum implementations

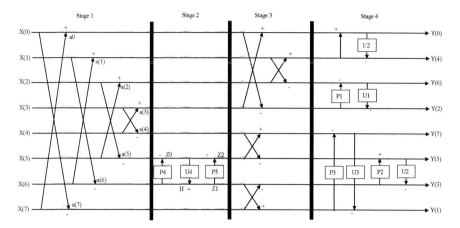

Fig. 2 Structure of the BinDCT

of arithmetic operations. Each one of these configurations has a dissimilar approximation of "P" and "U" values and a different number of addition and shifts. In this chapter, the configuration called C7 has been exploited as mentioned in Fig. 2.

4 Proposed Reversible BinDCT Design

The used configuration of the BinDCT, which is proposed in our previous work [42], is composed of four stages. This architecture requires 28 additions/subtractions and 28 shift operations. In this section, we discuss in detail the reversible BinDCT architecture

4.1 Study of the First Stage

In the first stage, we use Fan-out generator circuit (FG), 8-bit parallel adder, and 8-bit parallel subtractor as shown in Fig. 3.

Fan-Out Generator Circuit (FG)

In the first stage, the role of Fan-out generator sub-module (FG) is used to create replications of the eight inputs line X0–X7 (8-bit each line). Here, Feynman gate is used as Fan-out generation by setting zero to the B input. Figure 4a shows the proposed circuit which is composed of eight Feynman gates.

Corresponding simulation result of this sub-circuit is depicted in Fig. 4b. Hence, the total quantum cost of the Fan-out generator circuit is 8. The critical path for this circuit contains one FG gate ($\Delta FG = 1\Delta$).

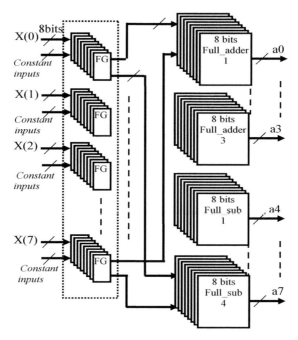

Fig. 3 Architecture of the stage 1

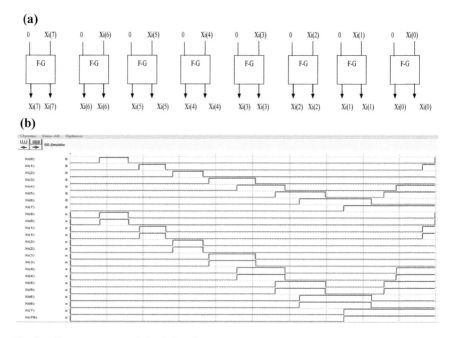

Fig. 4 **a** Fan-out generator, **b** its timing diagram

8-bit Full-Adder/Subtractor

Recently, various methods have been utilized to perform reversible adder/subtractor circuits [42, 43]. Here, we present a modified version of adder design presented in [42]. The designing of this sub-module can be accomplished with the use of PG gate and FG gate. Figure 5a shows the implementation of the designed adder/subtractor with the timing diagram in Fig. 5b. The reversible adder/subtractor executes a subtraction or addition according to the control line (Ctrl). The QC of the designed circuit is 9. The critical path for this circuit contains one FG and two PG gates. An 8-bit adder/subtractor can easily be constructed by cascading eight reversible adders/subtractors as shown in Fig. 6. The total QC of this module is 72, the number of garbage outputs is 16, and Depth = 72.

As can be seen from Fig. 3 that eight Fan-out generator circuits, four 8-bit full-adders, and four 8-bit full subtractors have been used. Consequently, this stage needs, respectively, 72 constant inputs, 128 GOs, and it has a QC of 612. The delay of the first stage = delay of fan-out circuit + delay of 8-bit adder/subtractor = $1\Delta + 72\Delta$ = 73Δ.

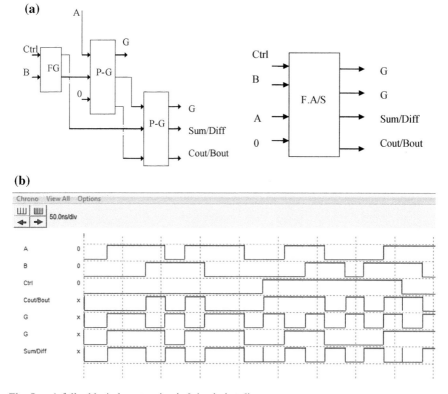

Fig. 5 **a** A full-adder/subtractor circuit, **b** its timing diagram

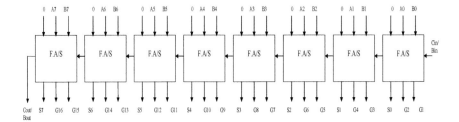

Fig. 6 Eight-bit full-adder/subtractors

4.2 Etude of the Second Stage

This stage is formed of four Fan-out generator circuits (Fig. 5), two 8-bit Full-Adders, two 8-bit Full-Subtractors, and four 8-bit universal shift registers (USRs). The Full-adder/subtractor design is proposed in the first stage (Fig. 6). The schematic of stage two is shown in Fig. 7 where

- $P4 = 1/2$
- $Z0 = a5 - 1/2 \times a6$
- $U4 = 3/4 = 1/2 + 1/4$
- $P5 = 1/2$

Reversible Universal Shift Register (USR)

The 8-bit USR is a register, which is composed of eight 4:1 multiplexers to drive the input signals of eight flip-flops in the register which are also connected to the clock input.

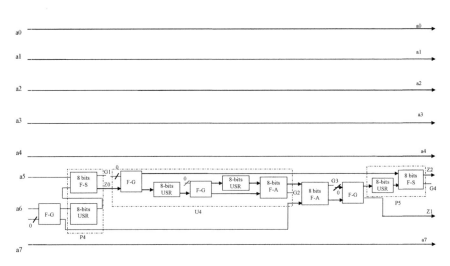

Fig. 7 Structure of the stage 2

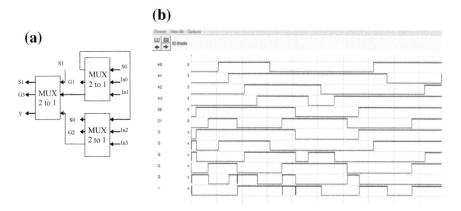

Fig. 8 **a** Proposed 4-to-1 multiplexer circuit, **b** its timing diagram

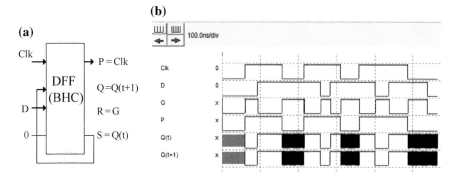

Fig. 9 **a** Proposed D Flip-Flop, **b** timing diagram

The proposed design of 4:1 reversible multiplexer is performed using three BHA gates [25] as depicted in Fig. 8a. The corresponding timing diagram of the designed 4:1 multiplexer is illustrated in Fig. 8b. According to Fig. 8a, we find that this design generates five garbage outputs and zero constant input. The QC of this structure is 12. The delay of the 4:1 multiplexer circuit is 8Δ. On the other hand, the realization of D Flip-Flop can be done using single BHC gate [25]. The output of D Flip-Flop is computed as follows:

$$Q(t+1) = Clk'.Q(t) + Clk.D \tag{1}$$

The structure and the timing diagram of this design are presented in Fig. (9a, b). The block diagram of the designed universal shift register is depicted in Fig. 10. Table 2 presents the truth table of the USR. The quantum cost of the proposed 8-bit universal shift register (USR) is 158. It has 30 ancilla inputs and 57 GOs. The delay of the proposed USR is $\Delta MUX + \Delta DFF + \Delta FG = 8 + 5 + 1 = 14\Delta$.

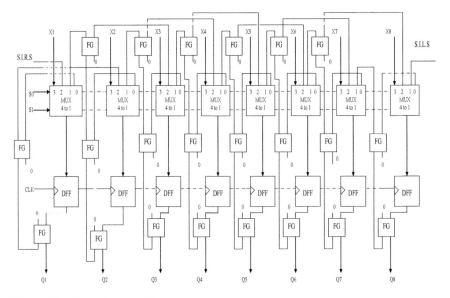

Fig. 10 Eight bits universal shift register

Table 2 Truth table of the USR

S1	S0	Final output
0	0	No change
0	1	Shift right
1	0	Shift left
1	1	Parallel load

Therefore, the total quantum cost (QC) of this stage is 952, 292 garbage outputs, and 184 constant inputs. The critical path of this stage is 348Δ.

4.3 Etude of Stage 3

The stage 3 is formed of 8 fan-out generator circuits, four 8-bit full-adders, and four 8-bit full-subtractors as depicted in Fig. 11. The same F-G circuit, USR, and 8-bit Full-Adder/Subtractor circuit in stage 1 and 2 are used in this stage. Clearly that stage 3 generates a total number of 128 CIs, 128 GOs. The total QC of this stage is 640. It has a delay of 73Δ.

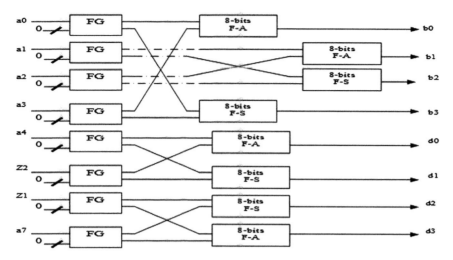

Fig. 11 Architecture of the stage 3

4.4 *Etude of Stage 4*

The stage 4 requires ten FG circuits, two 8-bits parallel adder, six 8-bits parallel subtractors, and eight USR as illustrated in Fig. 12. It generates a total number of 384 CIs and 584 GOs. The total QC of stage 4 is 1920. It has a delay of 174Δ.

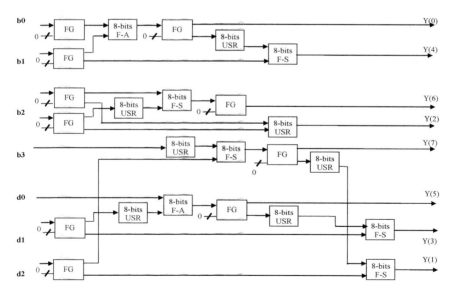

Fig. 12 Structure of the stage 4

5   Comparative Analysis and Simulation Results

In this work, the Microwind DSCH 3.5 tool is used to verify the functionality of
the designed reversible circuits [44] to check the correctness of the proposed sub-
modules. The simulation results indicate that the outputs of the designed sub-modules
are correctly obtained.

Tables 3, 4, 5, 6, and 7 demonstrate the comparative results between our proposed
sub-modules and other existing sub-modules. We find that the proposed sub-modules
of reversible BinDCT design outperform the existing ones.

Table 3 Comparison of various 1-bit full-adder designs

Design	Garbage output	Constant input	Quantum cost	Delay	Power(μW)
[42]	5	3	21	21Δ	N.A
[25]	4	2	18	N.A	13.7
[45]	1	2	14	8Δ	N.A
[46]	5	3	19	19Δ	N.A
[42]	3	1	14	14Δ	N.A
[42]	3	1	10	10Δ	N.A
Proposed design	2	1	9	9Δ	11.29

Table 4 Comparison of various 8-bit full-adder designs

Design	Garbage outputs	Constant inputs	Quantum cost	Delay	Power (μW)
[42]	42	34	192	N.A	N.A
[42]	48	34	168	N.A	N.A
[34]	24	16	138	N.A	N.A
[25]	25	16	124	N.A	204
[42]	23	8	76	N.A	N.A
Proposed design	16	8	72	72Δ	172.64

Table 5 Comparison of 4-to-1 reversible multiplexer designs

Design	Number of gates	Garbage outputs	Quantum cost	Delay	Power (μW)
[47]	3	5	15	15Δ	N.A
Proposed design	3	5	12	12Δ	22.3

Table 6 Comparison of reversible D-Latch

Design	Number of gates	Garbage outputs	Quantum cost	Delay	Power(μW)
[47]	2	2	7	7Δ	N.A
[48]	2	2	6	6Δ	N.A
Proposed design	1	2	5	5Δ	7.16

Table 7 Comparison of proposed of universal shift register with previous works

Design	Garbage outputs	Constant inputs	Quantum cost	Delay	Power(μW)
[48]	55	49	288	N.A	N.A
Proposed design	57	30	158	14Δ	269.33

6 Conclusion

Designing a fast and highly efficient architecture based on BinDCT algorithm is of great importance for low-power image and video compression systems. Reversible logic is the way to future computing technologies. It can be become obligatory because of the necessity to reduce power consumption better than the classical circuits. In this chapter, a novel design of BinDCT module based on reversible logic is obtained by combining the different implementations of reversible sub-modules, which is first ever proposed in the literature. The proposed circuit can be used in image and video compression in order, to meet the real-time constraints particularly on mobile devices. Future extensions such as various applications based on this reversible BinDCT module could be investigated.

References

1. Landauer R (1961) Irreversibility and heat generation in computing process. IBM J Res Dev 183–191
2. Bennett CH (1973) Logical reversibility of computation IBM J Res Dev 525–532
3. Shende VV, Prasad AK, Markov IL (2003) Synthesis of reversible logic circuits. IEEE Trans CAD 710–722
4. Knill E, Laflamme R, Milburn GJ (2001) A scheme for efficient quantum computation with linear optics. Nature 46–52
5. Chandana S, Navya C, Nagamani AN (2016) Design of register file using re versible logic. IEEE Int Conf Circuit Power Comput Technol (ICCPCT)
6. Chenga CS, Singh AK, Gopala L (2015) Efficient three variables reversible logic synthesis using mixed polarity Toffoli gate. Procedia Comput Sci 362–368
7. Maslov D, Dueck GW, Miller M (2008) Quantum circuit simplification and level compaction. IEEE Trans Comput-Aided Des. Integr. Circuits Syst 436–444

8. Saeedi M, Markov IL (2014) Synthesis and optimization of reversible circuits a survey. ACM Comput Surv (CSUR) 1–34
9. Cheng CS, Singh AK (2015) Heuristic synthesis of reversible logic–a comparative study. Adv Electr Electron Eng 210–225
10. Babu HM, Islam MR, Chowdhury AR, Chowdhury SMA (2004) Synthesis of full-adder circuit using reversible logic. In: 17th International conference on VLSI design, pp 757–760
11. Thapliyal H, Ranganathan N (2013) Design of efficient reversible logic-based binary and BCD adder circuits. ACM J Emerg Technol Comput Syst 1–31
12. Biswas AK, Hasan MM, Chowdhury AR, Babu HMH. (2008) Efficient approaches for designing reversible binary coded decimal adders. Microelectron J 1693–1703
13. Nagamani A, Ashwin S, Vinod KA (2014) Design of optimized reversible binary adder/subtractor and BCD adder. In: International conference on contemporary computing and informatics (IC3I)
14. Moghadam MZ, Navi K (2012) Ultra-area-efficient reversible multiplier. Microelectron. J 377–385
15. Morrison M, Ranganathan N (2011) Design of a reversible ALU based on novel programmable reversible logic gate structures. In: IEEE computer society annual symposium on VLSI
16. Rangaraju HG, Hegde V, Raja KB, Muralidhara KN (2012) Design of efficient reversible binary comparator. In: International conference on communication technology and system design, pp 897–904
17. Morrison M, Lewandowski M, Ranganathan N (2012) Design of a tree-based comparator and memory unit based on a novel reversible logic structure. In: IEEE Computer Society Annual Symposium on VLSI, pp 331–336
18. Antonino T, Matteo M, Gianluca P, Fabrizio F, Donatella S (2007) Pipelined fast 2D-DCT accelerator for FPGA-based SoCs. In: IEEE Computer Society Annual Symposium on VLSI, pp 9–11
19. Primechaev S, Frolov A, Simak B (2007) Scene changedetection using DCT features in transform domain videoindexing. In: 14th International workshop systems, signals and image processing and 6th EURASIP conference focused on speech and image processing, multi media communications and services, pp 369–372
20. Murphy C, Harvey M (2002) Reconfigurable hardware implementation of BinDCT. Electron Lett 1012–1013
21. Liang J, Tran T (2001) Fast multiplierless approximations of the DCT with the lifting scheme. IEEE Trans Signal Process 3032–3044
22. Philip PD, Paul MC, Truong QN (2005) BinDCT and its efficient VLSI Architectures for real-time embedded applications. J Imaging Sci Technol 124–137
23. Mahmoud FK (2007) Image compression using BinDCT for dynamic hardware FPGA's, thesis. Liverpool John Moores University
24. Abdessalem BA, Ichraf C, Abdellatif M (2016) Efficient BinDCT hardware architecture exploration and implementation on FPGA. J Adv Res 909–922
25. Lamjed T, Bouraoui O (2017) Design of hardware RGB to HMMD converter based on reversible logic. IET Image Process 646–655
26. Bikash D, Jadav CD, Debashis D (2017) Reversible logic-based image ste ganography using quantum dot cellular automata for secure nanocommunication. IET Circuits Devices Syst 1–10
27. Toffoli T (1980) Reversible computing. Technical memo MIT/LCS/TM-151, MIT lab for computer science
28. Feynman RP (1985) Quantum mechanical computers. Opt News 11–20
29. Chanderkanta AB, Santosh K (2017) Ultrafast optical reversible double Feynman logic gate using electro-optic effect in lithium-niobate based Mach Zehnder interferometers. In: Proceeding of SPIE, oxide-based materials and devices VIII, vol 10105, pp 1010520
30. Moraga C (2014) Mixed polarity reversible peresgates. IET Electron. Lett 987–989
31. Kaye P, Laflamme R, Mosca M (2007) An introduction to quantum computing Oxford University Press, Oxford, eBook-LinG, ISBN 0-19-857000-7
32. Fredkin E, Toffoli T (1982) Conservative logic. Int J Theoreical Phys 219–253

33. Mohammadi M, Eshghi M, Haghparast M, Bahrololoom A (2008) Design and optimization of reversible BCD adder/subtractor circuit for quantum and nanotechnology based systems. World Appl Sci J 787–792

34. Haghparast M, Navi K (2008) A novel reversible BCD adder for nanotechnology based systems. Am J Appl Sci 282–288; Peres A (1985) Reversible logic and quantum computers. Phys Rev A 3266–3276

35. Ali NB, Sajjad W, Nazir H (2015) A new approach of presenting reversible logic gate in nanoscale. SpringerPlus 153

36. Guowu Y, Hung WNN, Xiaoyu S (2005) Majority-based reversible logic gates. Theor. Comput. Sci, 259–274

37. Lenin G, Nor S, Mohd M (2014) Design and synthesis of reversible arithmetic and logic unit (ALU). In: IEEE conference computer, communications, and control technology (I4CT)

38. Stolze J, Suter D (2004) Quantum computing: a short course from theory to experiment. Wiley, Weinheim

39. Bruce J, Thornton M, Shivakumaraiah L, Kokate P, Li X (2002) Efficient adder circuits based on a conservative reversible logic gate. In: Proceedings IEEE computer society annual symposium on VLSI, pp 74–79

40. Murphy C, Harvey M (2002) Reconfigurable hardware implementation BinDCT. Electron Lett 1012–1013

41. Rangaraju HG, Venugopal U, Muralidhara KN, Raja KB (2010) Low power reversible parallel binary adder/subtractor. Int J VLSI Des Commun Syst 23–34

42. Shekoofeh M, Mohammad R, Reshadinezhad (2015) A Novel 4x4 Universal reversible gate as a cost efficient full adder/subtractor in terms of re versible and quantum metrics. Int J Mod Educ Comput Sci 28–34

43. Microwind DSCH—schematic editor and digital simulator. http://www.microwind.net/dsch.ph

44. Thersesal T, Sathish K, Aswinkumor R (2015) A new design of optical reversible adder and subtractor using MZI. Int J Sci Res Publ 1–6

45. Gupta A, Singla P, Gupta J, Maheshwari N (2013) An improved structure of reversible adder and subtractor. Int J Electron Comput Sci Eng 712–718

46. Shamsujjoha MH, Hasan M, Lafifa J (2013) Design of a compact reversible fault tolerant field programmable gate array: a novel approach in reversible logic synthesis. Microelectron J 519–537

47. Nazma TH, Hasan, BM, Lafifa J (2017) Power efficient optimum design of the reversible plessey logic block of a field-programmable gate array. J Sustain Comput 1–35

48. Dastan F, Haghparas M (2012) A novel nanometric reversible signed divider with overflow checking capability. Res J Appl Sci Eng Technol 535–543

Novel Approaches for Designing Reversible Counters

T. N. Sasamal, H. M. Gaur, A. K. Singh and A. Mohan

Abstract Reversible logic offers an alternative computation for future low-power computing devices. In this paper, an efficient and potent universal 33 and 44 reversible gates are considered to implement 4-bit counter. Performance of the proposed 33 gate is verified using thirteen standard three variables Boolean functions, which demonstrate from 17.8 to 45.2% superiority in term of gate counts obtained with other reversible gates. New structures for T flip-flop and D flip-flop, which utilize two efficient reversible gates are presented. These flip-flops and some existing gates are utilized to implement the Mod-16 counter and 4-bit Up/down counter. The reported architectures are modeled using VHDL and functional simulations are done using ISIM.

1 Introduction

Reversible logic offers a new computational paradigm for low power and high-speed nano scale device. As stated by Landauer [1], there will be at least $K_B T \ln 2$ Joules of heat dissipated out of the system for one bit of information erasure during computation. Bennett [2] showed dissipation close to zero is attendable, if information

T. N. Sasamal (✉)
Department of Electronics & Communication Engineering, NIT Kurukshetra,
Kurukshetra, India
e-mail: tnsasamal.ece@nitkkr.ac.in

H. M. Gaur
Department of Electronics & Communication Engineering, ABES Institute of Technology,
Ghaziabad (Delhi NCR), India
e-mail: leoharimohan84@gmail.com

A. K. Singh
Department of Computer Applications, NIT Kurukshetra, Kurukshetra, India
e-mail: ashutosh@nitkkr.ac.in

A. Mohan
Department of Electronics Engineering, IIT BHU, Varanasi, India
e-mail: profanandmohan@gmail.com

© Springer Nature Singapore Pte Ltd. 2020
A. K. Singh et al. (eds.), *Design and Testing of Reversible Logic*, Lecture Notes
in Electrical Engineering 577, https://doi.org/10.1007/978-981-13-8821-7_3

processing are carried out without erasing the information, i.e, by utilizing reversible gates. Reversible gates are the basic constituents of the reversible logic which allow a bijective relationship between input and outputs. Whereas the traditional gates cannot recover the input states from the generated output states due to its inherent irreversibility property. In this aspect, reversible logic design plays an important role in development of nanotechnology. Various research works have been offered in the direction including reversible computing [3–8], but less work has been reported for reversible sequential circuits. This paved the way to design new efficient structure for counters. The counters are the essential components of a digital system. The counter outputs depends on current input and previous state of the system, which drives the counter through predefined states. This feature helps to compute the number of clock pulses, timer and frequency divider, etc. In this work, we have presented synchronous Mod-16 counter and Asynchronous Up/Down counters by utilizing reversible gates. The rest of the chapter is organised as follows: Sect. 2 provides quick introduction to various exiting reversible gates along with two efficient reversible gates. Section 3 introduces two novel structures for 4-bit counter and corresponding performance analysis. Section 4 showcases the functional simulation results. Section 5 concludes the presented work.

2   Novel Reversible Gate

Reversible logic design utilizes reversible gates as the fundamental blocks. Reversible gates result bijective mapping between input and the output vectors. Hence, the number of inputs and outputs are equal. Several efforts have been made for efficient reversible gate designs [9–16]. Figure 1 depicts schematic of some of the primitive 3×3 reversible gates.

In [17], authors reported a universal parity preserving 4×4 reversible gate. It maps inputs (A, B, C) to outputs ($P = A \oplus C \oplus D$, $Q = D \oplus BC$, $R = C$, $S = D \oplus B \oplus C \oplus BC$). Figure 2 depicts the quantum realization of corresponding gate. The QC is 10 and quantum gate count 6. The SMS gate is utilized to design T flip-flop. A novel 3×3 reversible gate is proposed in [18]. It consists of 3 inputs (A, B, C) and 3 outputs ($P = A \oplus B \oplus C$, $Q = A.C+B'.C'$, $R = A.C+B.C'$). The quantum implementation of the PRG gate is given in Fig. 3. The QC is 9 and quantum gate count 5. The PRG gate is utilized to design D flip-flop. To evaluate the efficacy of PRG with respect to the other primitive gates, the 13 standard 3-variable Boolean functions have been considered among the 256 possible functions. The GC needed to realize the standard functions considering existing reversible gates is depicted in Table 1. Comparison displays the suitability of PRG to implement various logical functions in term of gate count (GC).

Fig. 1 Block diagram of **a**
Toffoli [9] **b** Fredkin [10] **c**
Peres [11] **d** QCA1 [12] **e**
RUG [13] **f** RM [14]

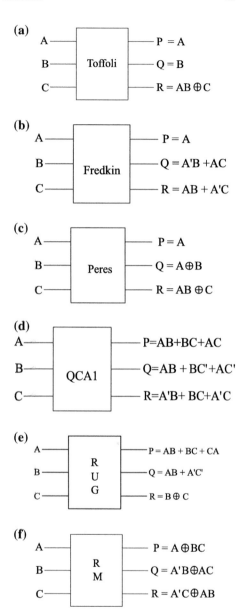

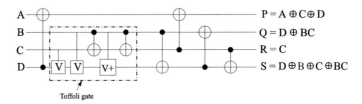

Toffoli gate

Fig. 2 Quantum realization of SMS gate

Fig. 3 Quantum
representation of PRG gate

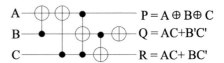

Table 1 Comparison of the 13 standard functions in terms of gate counts

Sr. No.	Functions	Toffoli	Peres	QCA1	Fredkin	RM	RUG	PRG
1	$F_1 = ABC$	2	2	2	2	2	3	2
2	$F_2 = AB$	1	1	1	1	1	1	1
3	$F_3 = ABC+A'B'C'$	3	3	2	3	2	2	3
4	$F_4 = ABC+AB'C'$	5	4	3	4	3	3	2
5	$F_5 = AB+BC$	2	2	2	2	2	3	2
6	$F_6 = AB+A'B'C$	5	3	3	5	2	3	2
7	$F_7 = A'BC+A'B'C'+ABC'$	6	4	3	6	3	3	3
8	$F_8 = A$	1	1	1	1	1	1	1
9	$F_9 = AB+BC+AC$	5	4	1	5	5	1	2
10	$F_{10} = AB+B'C$	3	3	3	1	1	3	1
11	$F_{11} = AB+BC+A'B'C'$	5	1	4	6	2	4	2
12	$F_{12} = AB+A'B'$	2	1	2	2	2	2	1
13	$F_{13} = ABC+A'B'C+$ $AB'C'+A'BC'$	2	3	2	3	2	2	1
Total		42	32	29	41	28	30	23
Improvement (%)		45.2%	28.1%	20.6%	43.9%	17.8%	23.3%	–

3 Reversible Counter

3.1 Traditional Counters

Conventional counters are broadly categories into two types: (1) Synchronous; (2) Asynchronous. In synchronous counters, all the flip-flops are activated simultaneously without adding delay incurred by individual flip-flop. Therefore the counter works at a higher frequency as compared to its counterpart ripple counter/

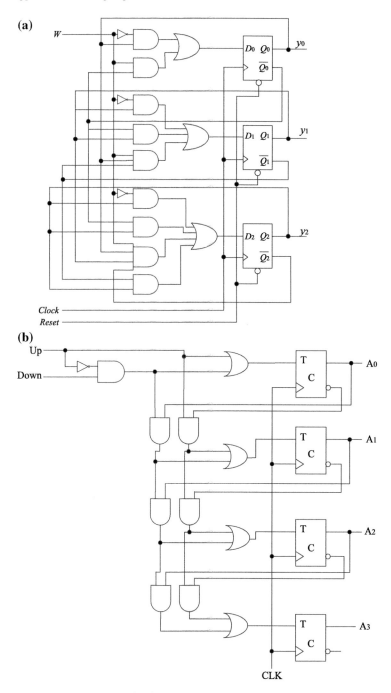

Fig. 4 Conventional counters **a** Mod-8 counter **b** 4-bit Up/Down counter

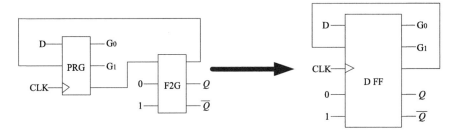

Fig. 5 Block diagram of reversible D flip-flop

asynchronous counter. A Mod-8 synchronous counter is depicted in Fig. 4a using D flip-flops [19]. An additional input signal W is provided for control purpose. If W input is 1, the circuit count is incremented else remains unchanged. Figure 4b illustrates conventional 4-bit synchronous Up/Down counter [20]. For the input Up = 1 or Down = 0, the circuit counts Up or Down, respectively (Fig. 4b).

3.2 Proposed Reversible Counter

3.2.1 Mod-16 Counter

The new reversible structure of the Mod-16 counter is presented in Fig. 6. As discussed in [19], the input D to the flip-flops can be expressed as follows:

$$D_0 = W \oplus Q_0$$
$$D_1 = W Q_0 \oplus Q_1$$
$$D_2 = W Q_0 Q_1 \oplus Q_2$$

The design utilizes four D flip-flops, 3 NOT gates, two Feynman gates (FG) along with three Toffoli gates ($3 \times 3, 4 \times 4, 5 \times 5$) to counts from 0 to 15. These Toffoli gates are responsible for generating the inputs D_1 to D_3, while one FG is used to produce D_0. The reversible D flip-flop block comprises the PRG gate and F2G) as depicted in Fig. 5. All the flip-flops are triggered with positive edge of the clock signal and generate outputs Q and Q'. The quantum implementation of the proposed Mod-16 counter requires quantum cost (QC) of 99, ancilla (constant) input of 12, and garbage output of 8. A n-bit version of the Mod-16 counter is implemented by cascading n D flip-flops, $(n - 1)$ NOT gates, $(n+1)$ Feynman gates, and $(n - 1)$ Toffoli gates with number of control lines $m \in 2, 3, \ldots, n$. For the n-bit counter, the QC is $(11 \times n) + (n - 1) \times 1 + (n + 1) \times 1 + \sum (2^{m+1} - 3), m \in 2, 3, \ldots, n$; i.e., QC $= 13n + \sum (2^{m+1} - 3), m \in 2, 3, \ldots, n$; $13n$ constant inputs, and $2n$ garbage outputs.

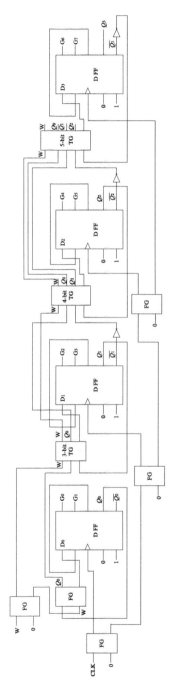

Fig. 6 Reversible implementation of Mod-16 counter (synchronous)

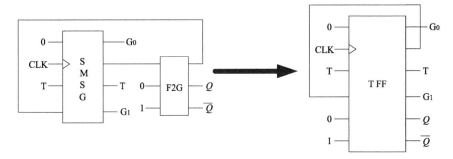

Fig. 7 Block diagram of reversible T flip-flop

Table 2 Performance comparison of different type of reversible 4-bit Up/Down counters

	QC	Ancilla input	Garbage output
Reference [21]	96	23	20
Presented circuit	104	22	17

3.2.2   4-bit Up/Down Counter

A new reversible structure for the 4-bit Up/Down counter is presented as in Fig. 8. This design utilizes four T flip-flops, 6 Peres gates, one Feynman gate (FG) along with two Modified Toffoli gates to counts from 0 to 15. The reversible T flip-flop block comprises the SMS gate and F2G as shown in Fig. 7. All the flip-flops are triggered with positive edge of the clock signal and generate outputs Q and Q'. The performance of reversible 4-bit controlled Up/Down synchronous counter can be illustrated easily after referring to the results in Table 2. The quantum implementation of the proposed 4-bit Up/Down counter requires quantum cost (QC) of 104, ancilla input of 22, and garbage output of 17. A n-bit version of the 4-bit Up/Down counter is designed by cascading n T flip-flops, $(n-1)$ modified Toffoli gates, one Feynman gate, and $2(n-1)$ Toffoli gates. For the n-bit counter, the QC is $(12 \times n) + 2(n-1) \times 4 + (n-1) \times 5$, i.e., $QC = 25n - 13$, $6n - 2$ constant inputs, and $4n + 1$ garbage outputs.

4   Functional Verification

Proposed circuits are modeled in VHDL and using Xilinx ISE 14.7. To check the logic functionality of the presented structures, functional simulations have been performed on ISim simulator. Simulation result of the reversible Mod-16 counter is depicted in Fig. 9. In Fig. 9, when $W = 1$, the counter counts from 0 to 15 and return to 0 on every positive edge of the clock signal. Simulation results of the proposed 4-bit Up/Down

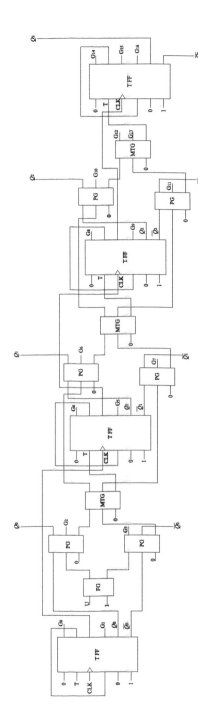

Fig. 8 Reversible implementation of 4-bit Up/Down counter (Asynchronous)

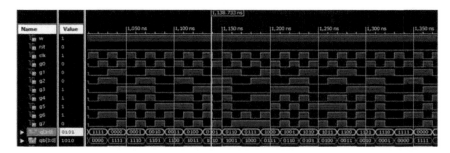

Fig. 9 Simulation result of reversible Mod-16 counter ($W = 1$)

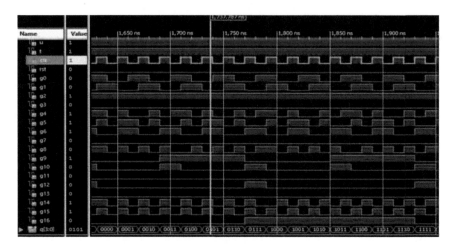

Fig. 10 Simulation result of reversible 4-bit Up counter ($U = 1$)

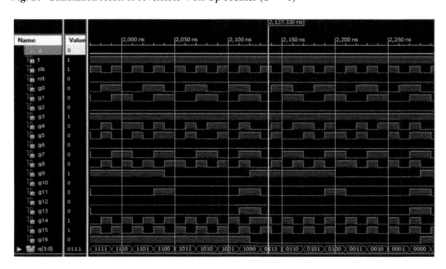

Fig. 11 Simulation result of reversible 4-bit Down counter ($U = 0$)

counter is presented for control input $U = 1$ and $U = 0$ as shown in Figs. 10 and 11, respectively. For $U = 1$, the counter works as a Up counter and counts from 0 to 15 and back to 0 on every rising edge of the clock signal. For $U = 0$, the counter behaves as a Up counter which counts from 15 to 0 and back to 15 on every positive edge of the clock signal. All these results reveal the correct logical functionality of the proposed designs.

5 Summary

This work shows efficient designs of reversible Mod-16 counter and 4-bit Up/Down counter. A new structure of Mod-16 counter is proposed using D flip-flops, which are based on a compact 3×3 reversible gate. Whereas the presented 4-bit Up/Down counter uses T flip-flops, which are based on a area efficient 4×4 reversible gate. The performances of the designs are evaluated considering quantum cost, garbage output, and ancilla inputs. All the designs are modeled using VHDL, while functional verifications are done using ISIM simulator. To show universality, n-bit counters are also demonstrated.

References

1. Landauer R (1961) Irreversibility and heat generation in the computational process. IBM J Res Dev 5:183–191
2. Bennett CH (1973) Logical reversibility of computation. IBM J Res Dev 17(6):525–532
3. Ren J, Semenov VK (2011) Progress with physically and logically reversible superconducting digital circuits. IEEE Trans Appl Supercond 21(3):780–786
4. Knil E, Laflamme R, Milburn GJ (2001) A scheme for efficient quantum computation with linear optics. Nature 46–52
5. Sasamal TN, Singh AK, Mohan A (2016) Efficient design of reversible ALU in quantum-dot cellular automata. Int J Light Electron Opt 127(15):6172–6182
6. Sasamal TN, Singh AK, Mohan A (2015) Design of two-rail checker using a new parity preserving reversible logic gate. Int J Comput Theory Eng 7(4)
7. Das JC, De D (2016) Novel low power reversible binary incrementer design using quantum-dot cellular automata. Microprocess Microsyst 42:10–23
8. Chabi AM, Roohi A, Khademolhosseini H, Sheikhfaal S (2017) Towards ultra-efficient QCA reversible circuits. Microprocess Microsyst 49:127–138
9. Toffoli T (1980) Reversible computing. Tech memo MIT/LCS/TM-151, MIT lab for computer science
10. Fredkin F, Toffoli T (2002) Conservative logic. Springer, Berlin
11. Peres A (1985) Reversible logic and quantum computers. Phys Rev A 32(6):3266
12. Ma X, Huang J, Metra C, Lombardi F (2008) Reversible gates and testability of one dimensional arrays of molecular QCA. J Electron Test 24:297–311
13. Sen B, Saran D, Saha M, Sikdar BK (2011) Synthesis of reversible universal logic around QCA with online testability. In: International symposium on electronic system design (ISED)
14. Sen B, Dutta M, Goswami M, Sikdar BK (2014) Modular design of testable reversible ALU by QCA multiplexer with increase in programmability. Microelectron. J. 45(11):1522–1532

15. Feynman RP (1986) Quantum mechanical computers. Found Phys 16(6):507–531
16. Parhami B (2006) Fault tolerant reversible circuits. In: Proceedings of the 40th asimolar conference on signals, systems, and computers (ACSSC), pp 1726–1729
17. Sasamal TN, Singh AK, Mohan A (2016) Design of parity preserving combinational circuits using reversible gate. In: 2nd international conference on next generation computing technologies (NGCT), Dehradun, pp 631–638. https://doi.org/10.1109/NGCT.2016.7877489
18. Sasamal TN, Singh AK, Mohan A (2018) Low complexity design of QCA reversible circuits via clock-zone-based crossover. Int J Theor Phys (IJTP) 57(10):3127–3140
19. Brown S, Vranesic Z (2008) Fundamentals of digital logic with VHDL Design, 3rd edn. McGraw-Hill Education, New York
20. Mano MM (2005) Digital design, 3rd edn. Prentice Hall, Englewood Cliffs
21. Rajmohan V, Ranganathan V (2011) Design of counters using reversible logic. In: 3rd international conference on electronics computer technology, vol 5, pp 138–142

Improving the Designs of ESOP-Based Reversible Circuits

C. Bandyopadhyay, S. Parekh, D. Roy and H. Rahaman

Abstract Finding ways to transform traditional circuits to energy efficient designs have found immense interest in present day's design industry and the quest of attaining low power consuming design techniques breeds the concept of reversible circuit design. One such computing paradigm that enforces reversibility in design architecture is quantum computation. Since a couple of years, several researches are going to make the reversible designs improved further and one such way of making reversible circuit efficient is to reduce the cost of the design. Aiming to build cost-efficient architectures, here in the work, we develop an improved synthesis approach for reversible circuit. The synthesis process has two phases. In the first phase, we present a best neighbour based circuit design scheme, where the functional outputs share the common data with their immediate neighbors and form shared designs. In the next phase, a circuit optimization algorithm runs and the circuits generated in previous phase pass through it for possible improvements in the design. The developed technique has been tested over different benchmark functions and the experiments show that the synthesis process followed by the optimization scheme substantially improves the design cost. To this extent, a comparative study with related existing techniques is undertaken and a brief analysis over the design approach is summarized at the end of the work.

C. Bandyopadhyay (✉) · H. Rahaman
Department of Information Technology, Indian Institute of Engineering Science and Technology, 711103 Shibpur, India
e-mail: chandanb@it.iiests.ac.in

H. Rahaman
e-mail: rahaman_h@it.iiests.ac.in

S. Parekh
Department of Engineering and Architecture, ID-Lab, University of Ghent, St. Pietersnieuwstraat 33, 9000 Ghent, Belgium
e-mail: shalini.parekh@ugent.be

D. Roy
Department of Computer Science, University of Central Florida, 32826 Orlando, FL, USA
e-mail: debashri@cs.ucf.edu

© Springer Nature Singapore Pte Ltd. 2020
A. K. Singh et al. (eds.), *Design and Testing of Reversible Logic*, Lecture Notes in Electrical Engineering 577, https://doi.org/10.1007/978-981-13-8821-7_4

1 Introduction

In last couple of years, the demand for energy efficient circuits and developing technologies for such architecture has received wide attention before the design industry.
In this conjuncture, "Reversible Circuit" [1, 2] has made a prominent footprint in the
industry as it promises to build low power consuming circuits. One such computing
platform that supports the design of reversible circuit is Quantum computing [3] and
it possible only due to the quantum mechanical property [4, 5] that enforces each
quantum operation to be reversible. The concept of reversible circuit is not restricted
to theoretical domain only as with the advancement of quantum circuit implementing technologies (like NMR [6], Superconducting qubit [6, 7], Ion Trap [8]) physical
implementation of reversible circuit has progressed [9].

In recent time, developing efficient synthesis methodology for reversible circuit
has gained much interest in research community, where areas like way for reducing
designs overheads (like quantum cost, gate count), making the circuit fault-tolerant
have come out as main design goals.

Though there exist variety of synthesis schemes, but some synthesis approaches
have proved to be very effective to form low cost-based designs. Depending on
the scalability level and the type of algorithm deployed, such approaches can be
distributed into two different classes.

1. **Optimal methods-based synthesis approach**: Optimal algorithms-based synthesis schemes are included in this category. Such schemes generally don't scale
   beyond 6-variable functions but within the said limit, it produces optimal solutions. Algorithms like Transformation technique [10], exact approaches [11]
   belong to this category.
2. **Sub-optimal solution-based synthesis scheme**: Solutions generated from such
   schemes never produce optimal results but such synthesis schemes can be scaled
   for large benchmark functions. Approaches like Genetic algorithms, A* [12], Ant
   Colony-based optimization technique [13], Simulated annealing-based synthesis
   process [14], Binary Decision Diagrams (BDD)-based design approach [15] and
   ESOP [16] -based representation schemes belong to this class.

Among these sub-optimal techniques, BDD and ESOP-based design approaches
are known to be effective techniques for synthesizing very large size functions.
But in comparison with BDD, ESOP generated circuits incur less design overhead
as such designs don't require huge ancille lines like BDD needs. Though several
improvements on ESOP designs have already been made in last couple of years, but
still remain scopes to improve it further.

Aiming to make efficient ESOP representation, here in our work, we present a best
neighbour based circuit design approach which first forms shared ESOP structure and
then optimize it by executing a template matching scheme. Resulting circuits from
this design approach show promising results compared to existing related works.

The rest of the chapter is organized as follows. Section 2 presents preliminaries
on reversible logic. The developed synthesis scheme is stated in Sect. 3. Section 4
presents experimental results and discussion. Finally, Sect. 5 concludes the work.

2 Background

Here, we introduce the preliminaries on reversible circuit and cost parameters used to evaluate efficiency of reversible circuits. A brief review on ESOP-based synthesis is also summarized at the end of this section.

2.1 Basics on Reversible Logic and ESOP-Based Design

Definition1 A circuit is termed "Reversible Circuit" if it holds the following four principles:

1. Circuit contains same number of input and output lines.
2. For each input pattern, there should exist a unique output pattern, i.e., mapping should be one to one.
3. The design of circuit will be such that no fan-out should exist in the design.
4. Circuit contains reversible gates only.

There exist a set of well-known reversible gates like NOT, CNOT, Toffoli [17], Feynman [18], Fredkin [19] which are widely used to form reversible circuits. These reversible gates act over a set of circuit lines, where the lines that contain the control nodes are known as control lines and function outputs are collected from target lines.

In the next example, we discuss the construction of reversible using reversible gates.

Example1 In Fig. 1a, the design of a reversible gate is shown. The depicted gate has two control nodes on line x_1 and x_2, and a target node on line t that implements the function $f = t \oplus x_1 x_2$. This gate is known as Toffoli gate. Like as Toffoli, the design of CNOT and NOT gate also has been depicted in Fig. 1b, c.

When such reversible gates are appended over a set of control and target lines, then a reversible circuit is formed. The design of a reversible circuit is given in Fig. 1d,

Fig. 1 a Toffoli Gate,
b CNOT Gate, c NOT Gate,
d reversible circuit
containing 4 reversible gates

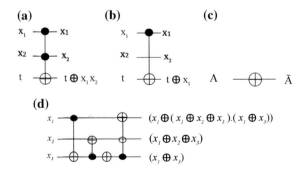

where the circuit contains four reversible gates incurring a cumulative design cost of 8.

As discussed, there exist several design approaches to construct reversible circuit and among those, ESOP is very widely known design technique. ESOP has a different representation format which is discussed next.

Definition2 Like as Sum of Products (SOP) expression, where product terms are distanced by ADD (+) operator, in ESOP, such terms are set apart using XOR ($\oplus$) operator. Any SOP expression can be transformed into an ESOP function by using a set of transformation rules. For expressing an n-input, m-output function using ESOP representation, it needs total $(n + m)$ number of circuit lines, where n represents number of existing control lines and the variable m stands for number of target lines or functional outputs in the design.

For example, if the set of transformation rules are executed over a certain logic function $f = xy + yz$ (of SOP form), then its equivalent ESOP form turned to $f = xy \oplus yz \oplus xyz$. But to form ESOP-based designs from benchmark files, an intermediate description is obtained and it is termed as *cubelist*.

Definition3 A cubelist is a collection of control node information expressed in a 2D matrix form $(m \times n)$, where each row represents a gate and the value in the cells of that row contains the control node information.

Example2 In Fig. 2a, a cubelist representing a benchmark function is given, where the cubelist is formed with five ESOP cubes. Once the cubes are mapped to respective reversible gates, an ESOP representation is obtained in Fig. 2b. While mapping the cubes to respective gates, the 1 bit values in the cubelist are converted to positive control nodes and the 0 bit values to NOT gates.

2.2 Cost Functions for Reversible Circuit

For evaluating the efficiency of reversible circuit, two cost matrices, namely quantum cost [20] and gate count are considered as prime driving factors.

Quantum Cost (QC): Minimum number of quantum operations involves to represent a quantum functionality into a circuital form is termed as quantum cost. The

Fig. 2 **a** Cube list for function f $(x_1, x_2, x_3, x_4) =$ $x_2 x_3 \oplus x_4 \oplus x_1 \oplus x_1 x_2 x_4$ $\oplus x_1 x_2 x_3 x_4$, **b** equivalent ESOP representation for the cube list of Fig. 2a

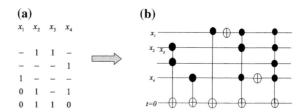

quantum cost of a reversible circuit is the cumulative cost sum of all the gates present in the design.

Gate Count (GC): The number of gates exists in a circuit is known as gate count of the design.

For, example a Toffoli gate has a standard cost of 5, CNOT and NOT each adds 1 cost to the design. The *gate count* metric for the circuit of Fig. 2b is 7 that incurs total quantum cost of 51.

Apart from the above-mentioned cost parameters, in recent time, three new cost matrices (T-count, T-depth [21] and NNC (nearest neighbor cost) [22]) have been introduced, which sometimes are used to judge the design efficacy.

2.3 Brief Review on ESOP Related Works

The way for transforming an arbitrary circuit to an equivalent ESOP design was first introduced [23], where the authors first claimed that to represent a n-input m-output function into ESOP form, it requires total $(2n + m)$ circuit lines but later they improved the design and bring down the line requirement to $(n + m)$ only.

By finding the structural symmetries between reversible gates using corelations, an improved ESOP synthesis technique has been developed in [24], where an auto-correlation coefficient-based cost function is also derived for identifying the gate positions in a network.

A greedy-based template matching technique for optimizing ESOP network has been introduced in [25], where the circuit costs are reduced considerably.

The technique for sharing of control lines to form shared structures followed by reordering of cubes to reduce NOT gate requirement has been introduced in [26]. But this approach has a limitation as the technique does not work effectively when common structural similarities between functional outputs are missing.

Further, the sharing concept in ESOP representation is improved in [27], where cube clustering-based design approach is developed. Though, in the design, authors have added a new circuit line for making the cluster feasible but the cost of the designs has been reduced considerably.

An improved way for finding better ESOP designs by forming shared representation is introduced in [28], where reordering of cubes for minimizing NOT gate usage has been undertaken. For further optimizing the designs, the functional power of negative control Toffoli gate is exploited in the design architectures. The method presented in [28] has shown very promising results for ESOP-based designs.

3 Proposed Technique

The circuits generated from synthesis algorithms are not always optimized as there exist several ways by which the designs of the circuits can be improved further. Such a way is sharing of common sub-expressions between output lines that turn a simple circuit into a shared design architecture. In our design methodology, we have derived such a strategy that not only forms shared designs but also eliminates redundant gates from the circuit.

This developed synthesis scheme is accomplished in two consecutive phases. In the first phase, a best neighbor-based circuit design scheme is introduced, where functional outputs first find the best neighbors and then share their functionalities with the selected neighbor. The circuits generated from this stage are termed as shared circuits. Though these shared circuits have cost-efficient representation but they may contain redundant gates in the design. So for improving the designs further, in the second phase, these shared circuits pass through a set of optimized templates where the sub-circuits in the shared designs are replaced with appropriate templates. In the experimental verification, it is seen that the best neighbor scheme followed by optimization procedure considerably improves the design by reducing the quantum cost. Now, here we are explaining both the design phases in detail.

Phase 1: Construction of shared ESOP designs

The primary objective of this design phase is to form *shared circuits* by employing the best neighbor strategy. Previously, we have defined that an output line can be termed as best neighbor for a functional output if it shares maximum common terms in-between, but finding a best neighbor involves some heuristic design policy. In our design scheme also, we have come up with a such heuristic strategy that first scans around a functional output and then select the best partner with whom later it shares the common functional terms. This strategy involves the following four steps.

1. Obtain a (*.esop*) file for a benchmark function using some existing cubelist generation tool (like EXORCISM-4 [29] that transforms a *.pla* file into an equivalent *.esop* file). Now, if all the cubes from this *.esop* file are mapped into respective gates over a set of circuit lines then an ESOP representation can be made.

2. Next, read the (.esop) file and generate all possible (n C 2) distinct combination pairs (variable n represents the number of functional outputs in the function) for the output lines, where we compute the number of common product terms present between the selected pair sets.

3. Hence after, prepare an information chart based on the common shared cubes of functional outputs and the list is sorted in such a way that the highest shared functional pairs appear at the beginning of the list. Now, a functional output is declared as the best neighbor of another functional output if it has shared maximum common terms in-between. But while finding the best matched partners, the choice for selected pairs should be such that no two pairs will contain the same functional outputs which are previously shared by another selected pair. This selection process generates (n C 2) number of distinct pairs from n output lines.

4. Now, the chosen selected pairs are allowed to share the common sub-expressions between, which finally make a shared ESOP representation. Now, we are explaining the developed approach using two different examples. In the first one, we

have considered a benchmark function containing even number of functional outputs, whereas in the second example, a function comprising with odd number of outputs. The reason for taking different functions is to show the effect of best neighbor pairs for odd and even number of functional outputs.

Example3 Consider the following benchmark function *"aj-e11_81"* as given in Fig. 3a. This function contains four input lines and four output lines. By following the process stated in step1, we obtain the .esop file corresponding to the input function by executing the source file (.pla file) through tool EXORCISM-4. Now, we initiate the method stated in step2 and generate ($^4C_2$) distinct combinations {(2,3), (2,4), (1,2), (1,3), (3,4), (1,4)} over the functional outputs. Next, prepare a sorted list containing the pair sets with their shared information. After the completion of step2, finally we obtain the information chart as shown in Fig. 3b.

Next, we initiate the scanning of the information chart in top-down fashion and mark the first output pair (2, 3). This output pair shares a total 6 cubes in-between second and third functional outputs and next we move the (2, 3) pair set in the final sharing table. Again, we scan the chart from where we stalled and find the next pair set (2, 4) but this output pair cannot be selected as the second functional output is existing in the pair set (2, 3), i.e., $(2, 3) \cap (2, 4) \neq \phi$.

Similarly, we refrain from selecting the next three output pairs as if they were selected then the mutual exclusion property will be violated. As there does not exist any common functional output between pair sets (2, 3) and (1, 4), so the selected pair (1, 4) is moved to final sharing table.

Once we reach the bottom of the information chart, we stop scanning and a final sharing table is formed in Fig. 3c. After performing the cube to gate mapping from the information of final sharing table, a shared design is obtained in Fig. 3d.

This design has a total cost of 128 but in spite of considering the best neighbor scheme if a simple circuit was formed from the cubelist of Fig. 3a then the design cost would be 218. So, it is very evident that the proposed scheme improves the design overhead by reducing quantum cost to a greater extent.

Next, we consider a function containing odd numbers of functional outputs and will check the application of best neighbor scheme over it.

Example4 Here, now consider the benchmark circuit *"pcler8_190"* containing 16 inputs and 5 functional outputs. As like example3, here also we obtain the .esop file from the benchmark's source file using tool EXORCISM-4 (see in Fig. 4a). After executing the step2 of our algorithm, an information sharing chart is prepared in Fig. 4b. In step3, we first select and then move the cube pairs (3.4) and (1,2) in final sharing table as depicted in Fig. 4c. After mapping each cube into respective gate and sharing the common terms as per the information generated in final sharing table, a complete design is obtained in Fig. 4d. In this problem also, as like Example 3, if a direct ESOP representation was made (from the cube list of Fig. 4a) then the design cost would be 368 but in our shared representation (Fig. 4d), the cost has upgraded to 320.

(a)

$x_1\ x_2\ x_3\ x_4\ f_1\ f_2\ f_3\ f_4$

```
–  –  0  –  0  1  1  1
0  1  0  –  1  1  1  1
–  0  0  0  1  1  1  1
–  0  –  1  1  1  1  0
–  –  1  1  0  1  0  1
0  –  1  1  1  1  1  1
1  –  –  0  1  1  0  0
0  –  –  1  0  1  1  1
–  1  –  –  1  0  1  0
```

(b)

line pairs	Cube's frequency over the o/p lines	Cube's identities					
2,3	6	1	2	3	4	6	8
2,4	6	1	2	3	5	6	8
3,4	5	1	2	3	6	8	
1,2	5	2	3	4	6	7	
1,3	5	2	3	4	6	9	
1,4	3	2	3	6			

(c)

Line pairs	No. of cubes shared	Cube's identities
2, 3	6	1 2 3 4 6 8
1, 4	3	2 3 6

(d)

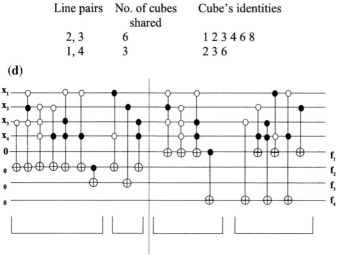

Fig. 3 **a** Cube list for benchmark function *aj-e11_81*, **b** possible functional output pairs and their sharing information, **c** computed sharing information chart, **d** shared ESOP representation of function *aj-e11_81*

Phase 2: Improving the best neighbor designs via circuit optimization

Though the best neighbor scheme effectively reduces the design cost but still remains scopes to improve the designs further. Aiming to re-optimize the circuits generated from previous stage, here we introduce a circuit optimization technique that not only brings down the quantum cost further but also reduces gate count from the designs. Basically, this optimization technique searches through the input circuit and replaces the structure matches with appropriate templates. This optimization process works in two steps, where in the initial step, a template matching-based optimization process executes and in the next step, the control lines in the resultant designs are shared in-between for making the designs efficient further.

(a)

X_1	X_2	X_3	X_4	X_5	X_6	X_7	X_8	X_9	X_{10}	X_{11}	X_{12}	X_{13}	X_{14}	X_{15}	X_{16}	f_1	f_2	f_3	f_4	f_5
–	–	1	1	1	1	–	–	–	–	–	–	–	–	–	–	1	0	0	0	0
–	–	1	1	1	1	–	–	–	0	–	0	–	–	–	–	0	0	1	1	0
–	–	1	1	1	1	–	–	–	0	–	–	–	–	–	–	0	1	0	0	0
–	–	1	1	1	1	–	–	–	0	–	0	1	–	–	–	0	0	0	1	0
–	–	–	–	1	1	–	–	–	0	–	0	1	–	–	–	1	0	0	0	0
–	–	–	1	–	–	–	–	1	–	1	–	–	–	1	1	0	0	0	0	1
–	–	–	–	1	1	–	–	–	–	–	–	–	0	–	–	0	0	0	1	0
–	–	–	–	0	1	–	1	–	–	–	–	–	–	–	–	0	0	0	1	0
–	–	–	–	–	–	–	–	–	–	–	–	–	–	–	–	1	1	1	1	0
–	–	–	–	1	1	–	–	–	–	–	–	0	–	–	–	0	0	1	0	0
–	–	–	–	0	1	1	–	–	–	–	–	–	–	–	–	0	0	1	0	0
–	–	–	–	1	1	–	–	–	–	–	0	–	–	–	–	0	1	0	0	0
–	1	–	–	0	1	–	–	–	–	–	–	–	–	–	–	0	1	0	0	0
1	–	–	–	0	1	–	–	–	–	–	–	–	–	–	–	1	0	0	0	0

(b)

line pairs	Cube's frequency over the o/p lines	Cube's identities				
3,4	2	2	9			
1,2	1	9				
1,3	1	9				
1,4	1	9				
2,3	1	9				
2,4	1	9				
1,5	0					
2,5	0					
3,5	0					
4,5	0					

(c)

Line pairs	No. of cubes shared	Cube's identities	
3, 4	2	2	9
1, 2	1	9	

Fig. 4 **a** The computed cubelist for benchmark circuit pcler8_190, **b** possible functional output pairs and their sharing information, **c** computed information sharing chart, **d** shared ESOP representation of function pcler8_190

(d)

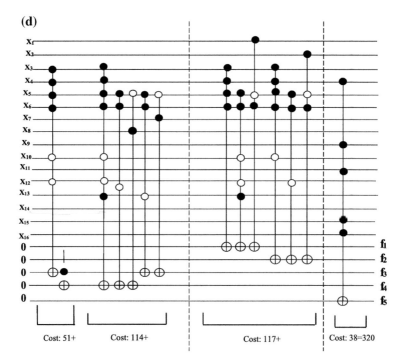

Fig. 4 (continued)

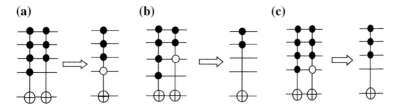

Fig. 5 The defined templates. **a** Proposed Template1, **b** Proposed Template2, **c** Proposed Template3

Step1: The templates as shown in Fig. 5 first form a template library and once a circuit passes through that library, the templates try to find their match from the input circuit. If a match is found then the sub-circuit is replaced with the selected template structure.

But if the templates as defined in Fig. 5 do not find any matches then the circuit passes through a set of new templates as depicted in Fig. 6, which again expand the internal design based on the *hamming distance* of gates and create a scope for re-optimization using the Fig. 5 templates. All the templates defined in Figs. 5 and 6 have very generalized structures as they can be employed over any size of circuit.

Step2: In this phase, the Phase1 generated circuit passes through a control node sharing rule, which first finds common control nodes in-between two suc-

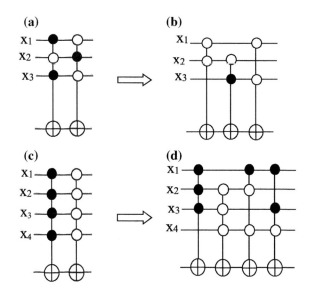

Fig. 6 **a** Two gates with 3 hamming distance apart, **b** transformed circuit from Fig. 6a, **c** two gates with 2 hamming distance, **d** transformed circuit from Fig. 6c

cessive gates and in later time forms a shared control structure out of the input circuit.

The way of forming shared control structures from input circuits is given in Fig. 7, where it can be seen that both the input circuits of Fig. 7a, c have initial cost of 26 and 42, but after reconfiguring the circuits by sharing the common controls, the costs have brought down to 15 and 25, respectively. The modified designs after the improvements are shown in Fig. 7b, d.

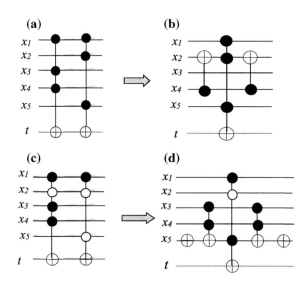

Fig. 7 **a** Initial design incurring total QC = 26, **b** QC comes down to 15 in after the sharing of common control nodes, **c** ESOP circuit incurring QC = 42 before sharing, **d** cost reduced to 25 after the sharing of common control nodes

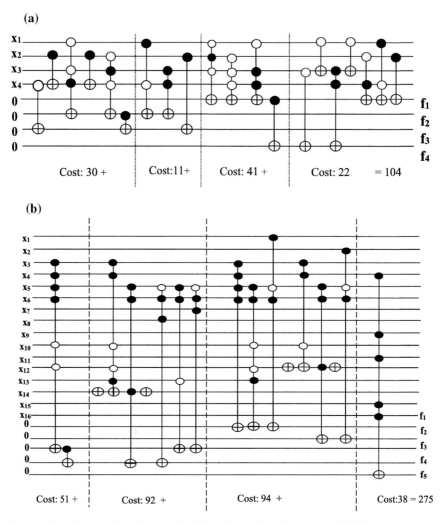

Fig. 8 **a** The obtained circuit from step2 of Phase2 over Fig. 3d, **b** improved ESOP representation of function *pcler8_190*

In order to explain the execution of Phase2 of optimization, we are considering the same circuit (depicted in Figs. 3d, 4d) obtained from Phase1 of Examples 3 and 4. The resulting improved designs from Phase2 are given in Fig. 8a, b, where the cost for both the circuits has reduced to 104 (from 117) and 275 (from 320), respectively.

So, from the above examples it can be concluded that Phase1 followed by Phase2 of optimization improves designs by diminishing the cost of circuits. Experimental evaluation and analysis over the proposed design approach are summarized next.

4 Experimental Results and Analysis

The proposed synthesis method has been tested over a wide range of benchmarks [30] and the obtained results are summarized in the following two tables—Tables 1 and 2. In Table 1, a comparison between our approach and existing ESOP techniques ([24, 25] and [16]) is summarized, where our approach generated circuits have registered considerable improvements in design cost.

Table 1 Comparative study with existing ESOP-based synthesis approaches

Function names	In/out	QC from [25], approach I	QC from [24], approach II	QC from [25], approach III	QC from [16], approach IV	Our technique QC after Phase1	QC after Phase2	ET (in second)
3_17_6	3/3	–	45	46	39	39	35	<1
4gt12_24	4/1	43	43	43	43	42	37	<1
4gt11_23	4/1	5	5	5	5	5	5	<1
4gt10_22	4/1	35	35	35	–	34	34	<1
4mod7	4/3	167	169	169	143	147	131	<1
f2	4/4	246	–	–	116	210	162	<1
4_49_7	4/4	201	222	222	119	143	103	<1
aj-e11_81	4/4	201	221	217	116	128	104	<1
wim	4/7	218	–	–	150	206	168	<1
dc1_142	4/7	454	–	–	201	244	172	<1
Ex2	5/1	160	–	–	118	143	114	<1
Ex3	5/1	97	–	–	73	76	51	<1
C17	5/2	105	81	81	–	83	82	<1
cm82a	5/3	143	–	–	–	128	98	<1
rd53_68	5/3	269	–	–	136	246	194	<1
squar5_206	5/8	465	–	–	–	350	331	<1
C7552_119	5/16	–	2015	2015	1123	1535	967	<1
con1	7/2	206	–	–	179	163	163	<1
rd73_69	7/3	1150		–	820	1120	711	<1
Z4_224	7/4	642	412	420	260	551	334	<1
Z4ml_225	7/4	642	–	–	260	551	334	<1
sqrt8	8/4	616	–	–	–	568	320	<1
misex1_178	8/7	1012	–	–	743	714	624	<1
dk27_146	9/9	252	–	–	245	245	221	<1
max46	9/1	4968	–	–	3239	4524	2588	<1
add6	12/7	5757	6751	6764	–	5518	3491	<1
alu1_94	12/8	239	216	216	–	198	172	<1

(continued)

Table 1 (continued)

Function names	In/out	QC from [25], approach I	QC from [24], approach II	QC from [25], approach III	QC from [16], approach IV	Our technique		
						QC after Phase1	QC after Phase2	ET (in second)
t481	16/1	236	–	–	192	229	183	<1
pcler8	16/5	340	–	–	–	319	275	<1
mux	21/1	815	–	–	–	800	800	<1
cordic	23/2	348779	–	–	98456	187582	69723	<1

"–" symbol (dash): signifies "Non availability of results" ET stands for: Execution Time

Table 2 Comparisons between ESOP and non-ESOP related

Benchmark specification		RMRLS technique [31]			RMS technique [31]			Proposed design approach			
Function's name	In/Out	QC	GC	ET	QC	GC	ET	QC after Phase1	QC after Phase2	GC	ET
decod24_10	2/4	55	11	497	19	7	<0.01	19	15	6	<0.01
mini-alu_84	4/2	173	21	495.61	248	36	<0.01	85	85	6	<0.01
rd53_68	5/3	–	–	>500.00	2646	221	0.14	246	194	18	<0.01
mod5adder_66	6/6	529	37	494.46	151	35	0.06	477	301	30	<0.01
rd73_69	7/3	–	–	>500.	20779	1344	1.93	1120	711	44	<0.01
rd84_70	8/4	–	–	>500.	8738	124	9.92	3554	2388	29	<0.01
cycle10_2_61	12/12	1435	26	491.87	1837	41	26.17	1788	1444	42	<0.01
plus63mod4096	12/12	–	–	>500.	4873	24	17.74	956	955	27	0.01
plus127mod8192	13/13	–	–	>500.	9131	25	57.16	1462	1461	27	0.01
plus63mod8192	13/13	–	–	>500.	9183	28	57.19	1356	1227	31	<0.01

In Table 2, comparison with non-ESOP-based techniques (Reed-Muller based [31] and Transformation based [32]) has been summarized, where also cost improvement is observed.

Though the presented approach makes the circuits efficient by reducing the design cost but it can be improved further. The areas for possible improvements and the drawbacks of the presented method are briefed next.

- Phase1 of our approach cannot be executed over an input circuit if the circuit contains single output only but those functions that contain higher number outputs show considerable improvement in design cost if Phase1 of design strategy is followed.
- Sometimes, while selecting the best neighbor pairs, we have faced conflicting situations when two consecutive pair contains same number of shared terms. In that case, we have selected the first one but this consideration in some cases has lead to higher cost designs. But this selection mechanism can be improved further if more exact heuristic is derived.

- The design approach has certain reservation over the level of scalability as it cannot generate circuit beyond 30 inputs and this restriction is due to the higher time consumption while forming the *sharing chart* in Phase1 of design.
- Another important feature related to our design approach is, the optimization strategy (as mentioned in Phase2) not only very useful over Phase1 generated circuits but it can be employed for re-optimizing circuits generated from different synthesis algorithms.

5 Conclusion

In this work, we have detailed a best neighbor-based circuit generation scheme which makes improved representations by making shared designs. Along with this synthesis technique, we have also introduced a template-based optimization procedure which further improves the design cost generated from the best neighbor technique. All the design phases have successfully tested over a wide range of benchmarks and the related experimental findings are stated in result tables. Comparisons with related synthesis techniques also have been included in the result tables, where improvements in our approach have been registered. Though this approach does not have very higher level of scalability but can be considered as an efficient synthesis process as it brings down the design cost of circuits to a greater extent.

References

1. Landauer R (1961) Irreversibility and heat generation in the computing process. IBM J Res Dev 5:183
2. Bennett CH (1973) Logical reversibility of computation. J Res Dev 17(6):525–532
3. Nielsen M, Chuang I (200) Quantum computation and quantum information. Cambridge University Press, Cambridge
4. Shor PW (1994) Algorithms for quantum computation: discrete logarithms and factoring. In: Foundations of Computer Science, pp 124–134
5. Grover LK (1996) A fast quantum mechanical algorithm for database search. In: Theory of Computing, pp 212–219
6. Veldhorst M et al (2014) A two qubit logic gate in silicon. Nature
7. Ghosh J et al (2013) High-fidelity CZ gate for resonator based superconducting quantum computers. Phys Rev A 87:022309
8. Barends R et al (2014) Logic gates at the surface code threshold: Superconducting qubits poised for fault-tolerant quantum computing. Nature 58:500–503
9. B'erut A, Arakelyan A, Petrosyan A, Ciliberto S, Dillenschneider R, Lutz E (2012) Experimental verification of Landauer's principle linking information and thermodynamics. Nature 483(3):187–189
10. Golubitsky O, Falconer SM, Maslov D (2010) Synthesis of the optimal 4-bit reversible circuits. In: Proceedings of the 47th design automation conference. ACM, pp 653–656
11. Grobe D, Wille R, Dueck GW, Drechsler R (2008) Ex act multiple-control toffoli network synthesis with sat techniques., IEEE Trans Comput-Aided

12. Datta K, Rathi G, Sengupta I, Rahaman H (2012) Synthesis of reversible circuits using heuristic search method. In: 2012 25th international conference on VLSI design (VLSID). IEEE, pp 328–333
13. Li M, Zheng Y, Hsiao MS, Huang C (2010) Reversible logic synthesis through ant colony optimization. In: Design, Automation & Test in Europe Conference & Exhibition (DATE). IEEE, pp 307–310
14. Abdessaied N, Soeken M, Dueck GW, Drechsler R (2015) Reversible circuit rewriting with simulated annealing. In: 2015 IFIP/IEEE international conference on very large scale integration (VLSI-SoC). IEEE, pp 286–291
15. Wille R, Drechsler R (2009) BDD-based synthesis of reversible logic for large functions. In: Design automation conference, pp 270–275
16. Bandyopadhyay C, Rahaman H, Drechesler R (2014) A cube pairing approach for synthesis of esop-based reversible circuit. In: IEEE 44th International Symposium on Multiple-Valued Logic (ISMVL)-2014, Germany, pp 109–114. https://doi.org/10.1109/ISMVL.2014.27
17. Toffoli T (1980) Reversible computing. In Tech. Memo-MIT/LCS/TM-151, MIT Lab for Comp Sci
18. Feynman R (1986) Quantum mechanical computers. In: Foundations of Physics, vol 16, pp 507–531. (Originally appeared in optics news, February 1985)
19. Golubitsky O, Falconer SM, Maslov D (2010) Synthesis of the optimal 4-bit reversible circuits. In: Proceedings of the 47th design automation conference. ACM, pp 653–656 2010
20. Maslov D (2002) Reversible logic synthesis benchmark page. http://www.cs.uvic.ca/dmaslov/
21. Miller D, Soeken M, Drechsler R (2014) Mapping NCV circuits to optimized clifford + t circuits. Revers Comput
22. Wille R, Lye A, Drechsler R (2014) Exact reordering of circuit lines for nearest neighbor quantum architectures. IEEE Trans Comput Aided Des Integr Circuits Syst
23. Fazel K, Thornton M, Rice JE (2007) ESOP based Toffoli gate cascade generation. In: PACRIM, pp 206–209
24. Rice JE, Suen V (2010) Using autocorrelation coefficient-based cost functions in ESOP-based Toffoli gate cascade generation. In: CCECE, pp 1–6
25. Rice JE, Nayeem NM (2011) Ordering techniques for ESOP-based Toffoli cascade generation. In: PacRim, pp 274–279
26. Nayeem NM, Rice JE (2011) A shared-cube approach to ESOP-based synthesis of reversible logic. Facta Univ Ser: Electron Energ 24:385–402
27. Datta K, Rathi G, Sengupta I, Rahaman H (2013) An improved reversible circuit synthesis approach using clustering of ESOP cubes. In: 18th Reed-Muller workshop
28. Jegier J, Kerntopf P (2017) Application of the maximum weighted matching to quantum cost reduction in reversible circuits. In: MIXDES 24, pp 224–228
29. Mishchenko A, Perkowski M (2001) Fast heuristic minimization of exclusive-sums-of-products. In: 6th Reed-Muller workshop, pp 242–250
30. Wille R, Grosse D, Teuver L, Dueck GW, Drechslere R (2008) Revlib: an online resources for reversible functions and reversible circuits. In: ISMVL
31. Gupta P, Agrawal A, Jha NK (2006) An algorithm for synthesis of reversible logic circuits. IEEE Trans CAD 25(11):2317–2330
32. Maslov D, Dueck GW, Miller DM (2007) Techniques for the synthesis of reversible toffoli networks. ACM Trans Des Autom Electron Syst 12(4)

Logic Synthesis for Reversible Circuits

M. Fujita

Abstract Automatic synthesis methods of reversible circuits are discussed in this chapter. First the basic characteristics of reversible circuits are reviewed, and the difference of logic synthesis methods for general CMOS circuits and reversible circuits are clarified. Then logic synthesis methods for reversible circuits based on exhaustive search, repetition of local circuit transformations, Binary Decision Diagram (BDD) based approaches, SAT-based methods, and hierarchical methods are presented in order. The chapter concludes with discussions on future research topics.

1 Introduction

There has been a significant progress in the development of quantum computers and their related technology in recent years and more than 100 quantum bits (Qbits) is now physically realizable. Although it is still far away from practical data processing with quantum computers, such as solving computation intensive problems of practical sizes, these small-scaled quantum computers are actually used to explore the architecture and algorithms for large problems. For academic use, such as teaching, experiment, and developing algorithms, such quantum computers are really used by many people, and many new results are to come up. Also, there are efforts to develop new programming languages for quantum computers such that characteristics of quantum computers are appropriately encoded into the design of the programming languages. In terms of implementations with logic circuits for quantum computers, reversible circuits give the base technology for the interests of most people.

Also, it is proved that reversible circuits should be used for low power operations, as the power dissipation can be independent from the underlying technology as long as irreversible logic is used [1]. Moreover, it is known that in order for the power not to be dissipated in a circuit, it must be built from reversible gates [2].

Due to these and other reasons, there are more and more justifications to consider circuits composed of only or mostly reversible gates. In this chapter, we discuss

M. Fujita (✉)
University of Tokyo, Tokyo, Japan
e-mail: fujita@ee.t.u-tokyo.ac.jp

© Springer Nature Singapore Pte Ltd. 2020
A. K. Singh et al. (eds.), *Design and Testing of Reversible Logic*, Lecture Notes in Electrical Engineering 577, https://doi.org/10.1007/978-981-13-8821-7_5

automatic logic synthesis techniques for such reversible combinational circuits. That is, a network of reversible combinational gates is automatically generated from a given reversible function as the specification function. As for sequential circuits, they can be a natural extension of the reversible combinational circuits discussed in the chapter.

2   Overall View of Reversible Circuit Synthesis

A combinational logic circuit is said to be reversible if it generates a unique output value for each input value. That is, if the logic functions implemented with the circuit are represented in truth tables, the output values are equivalently some permutations of the input values. This is a very strong constraint when synthesizing combinational reversible circuits, and so the ways to generate circuits are very different from the logic synthesis methods for normal circuits, such as the techniques implemented in the logic synthesis and verification tool ABC [3]. In general, the goal is to provide synthesis methodologies by which trade-offs among circuit quality, efficiency, and feasibility are explored automatically or semi-automatically targeting reversible circuits.

Such circuits take an important role in quantum computing, optical computing, nanotechnology and low-power CMOS designs. There are several ways from the viewpoint of the automatic synthesis of reversible circuits. There have been efforts to generate minimum size (in terms of numbers of reversible gates, numbers of signal lines, or some others) circuits. Here it is called "exact synthesis". Although it gives best quality in terms of the measure of the circuits, generally the time for synthesis becomes much longer than practical, as the circuit size increases. As a result, it can be applied only to small reversible functions, and the interests are just for their theoretical characteristics.

Due to that, there are approaches which try to transform the given circuit into the corresponding reversible one by going through various optimization and/or transformation processes. Here it is called "transformation based" method. Although it may be applied to larger circuit synthesis, generally speaking, the size of reversible circuits which can be automatically synthesis is still highly limited. One such approach to synthesize relatively large reversible circuits is to generate them from Binary Decision Diagram (BDD) [4] representation as the specification function. It is called "BDD based" method and can theoretically synthesize reversible circuits having more than 10,000,000 gates, as it is now easy to build BDD having more than 1,000,000 nodes. The quality of the synthesized circuits is, however, unnecessarily much larger, as BDD representation sometimes becomes much larger than the number of gates in the multi-level logic circuits. BDD is also used in conjunct with the exact method and the transformation based method, as BDD is a very efficient way to canonically represent logic functions. BDD and Boolean satisfiability checking (SAT) are the two common methods to check the equivalence of two logic functions, and so they can be used as tools to logically check the applicability of transformations.

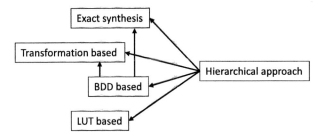

Fig. 1 Synthesis methods for reversible circuits

There are synthesis methods using LUT (Look Up Table) based approach. Here it is called "LUT based" method, and the given circuit is transformed into a LUT based network and then the corresponding reversible circuit is generated. This can be generalized as "hierarchical approach" and uses two stages, partitioning a given circuit in a set of synthesizable regions and synthesizing each region.

The above are illustrated as shown in Fig. 1. Basically the hierarchical approaches can be applied in conjunct with various synthesis methods which are applied to each resion after the partitioning, After the preparation and the introduction of basic definitions, they are discussed with some illustrative examples one by one in this chapter.

3 Preparation and Basic Definitions

For preparation, several definitions are given in this section for the discussions in the later sections. If necessary, additional definitions may be given in the later sections as well.

Definition 1 An m-input, m-output, Boolean function $f(X), X = (x_1, x_2, \ldots, x_m)$ is said to be reversible if it generates a unique output value for each input value.

Please note that in order to satisfy the above condition, the number of inputs and the number of outputs in a reversible function must be the same.

Definition 2 A function f is said to be linear if and only if the following is satisfied: $f(x \oplus y) = x(x) \oplus f(y)$, where $\oplus$ denotes bit-wise exclusive-OR operation,

A reversible function can be specified in a truth table. The mapping between inputs and outputs can be understood as an objective one. And so, the set of integers $0, 1, \ldots, 2^m - 1$ corresponding to the input values are mapped onto themselves as output values. Here the integers are recognized to be the values represented in the input values and output values as binary. Therefore, a reversible function can be considered as an ordered set of integers corresponding to the output values of the table with respect to the increasing order of input values. For example, 4, 6, 3, 1, 2, 0, 7, 5

Table 1 An example truth table

x_3	x_2	x_1	Integer	x_3'	x_2'	x_1'	Integer
0	0	0	0	1	1	0	4
0	0	1	1	1	0	0	6
0	1	0	2	0	1	1	3
0	1	1	3	0	0	1	1
1	0	0	4	0	0	0	2
1	0	1	5	0	1	0	0
1	1	0	6	1	1	1	7
1	1	1	7	1	0	1	5

represents the function shown in Table 1. Here the function over integers are interpreted as $f(0) = 4, f(1) = 6, f(2) = 3$, and so on.

Basically we can understand that a reversible function performs permutations in
the numbers corresponding to the input values and the output values.

Definition 3 An n-input, n-output gate is said to be reversible if it implements a
reversible function of n inputs and outputs.

Definition 4 An $n \times n$ Toffoli gate does not change the first $n - 1$ lines (used as
control), and complements the nth line (the target line), if the values of the control
lines are all 1 (i.e., their product is 1).

Definition 5 A permutation is called even if a product of an even number of trans-
positions (swapping two values) implements the permutation.

A number of different reversible gates have been proposed. Here we use Toffoli
gates defined as follows.

Definition 6 An $n \times n$ Toffoli gate is described as $TOF_n(x_1, x_2, \ldots, x_n)$ where x_n
is the target line. Here the prime symbol is used to denote the value of a line after
passing through the gate. Therefore the following are satisfied.

$$x_1' = x_1, \ x_2' = x_2, \ \ldots, \ x_{n-1}' = x_{n-1}, \ x_n' = x_1 x_2 \ldots x_{n-1} \oplus x_n$$

$TOF_1(x_1)$ is a special one where there is no control line, and x_1 is always complemented. That is, it corresponds to a NOT gate in general logic circuits. $TOF_2(x_1, x_2)$
is generally called controlled-NOT gate (CNOT). $TOF_1(x_1)$, $TOF_2(x_1, x_2)$, and
$TOF_3(x_1, x_2, x_3)$ are illustrated as shown in Fig. 2. Black dots show which are control
inputs and the white dots show the target line.

Definition 7 A SWAP gate exchanges its two inputs.

Fig. 2 Toffoli gates

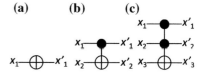

4   Exact Synthesis Method

First an exact synthesis method that works for circuits of even permutation on input values is shown as an introduction to the optimum synthesis for reversible circuits. It can be a base for the optimum synthesis from a given function through decomposition with branch and bound searches [5].

It can be shown that every function corresponding to a linear transformation, f can be realized only with CNOT gates. From this it can be concluded that every even permutation can be realized only with CNOT gates, and a branch and bounds method for exact synthesis based on this idea is shown in [5]. Although this is a very interesting result theoretically, it may not be applied to any realistic sizes of circuits, as the branch and bounds methods used here take long time in general.

More recently exact synthesis methods based on satisfiability (SAT) checking have been introduced [6, 7]. SAT checking is applied whether the current transformation is valid or not. Here, these methods are briefly discussed by showing small examples. For the details on the method, please see [6, 7]. Now we discuss the exact algorithms for the synthesis of networks consisting of multiple-control Toffoli gates, that is, automatic synthesis algorithms which guarantee that a network with the minimal number of gates is generated. They are based on iterative algorithms which treat the synthesis process as a repetition of generating general and parametrized circuits with a fixed number of Toffoli gates and its decision problem to see the generated circuit can implement the target reversible function or not. Those decision problems can be defined as Boolean satisfiability (SAT) or SAT modulo theory (SMT) problems. As soon as one of these decision problems is proved to be satisfiable, a network consisting of Toffoli gates corresponding to the given function has been identified as an instance of the generated general circuit. The performance of the synthesis, i.e., time for synthesis, highly depends on the encoding used in decision problems or how to represent the general and parametrized circuits. Generally speaking this is a time consuming process as SAT/SMT problems.

The basic idea of these exact synthesis methods is as follows: Given a reversible function f, the problem to generate a network consisting of Toffoli gates for the function is formulated as a sequence of generation of general and parametrized circuits and their associated Boolean decision problems which essentially check the existence of the implementation as an instance of the generated circuits. The formulation is to see whether there is a network consisting of exactly d Toffoli gates for the reversible function f and a given number d. This problem can be formulated as either a SAT or SMT problem and solved by the corresponding solvers.

For exact synthesis, the minimum value of d is to be found, as the circuits having minimum numbers of gates are the targets. A straightforward approach is to start searching a solution with the minimum value which is $d = 1$. If there is no solution, i.e., the decision problem is returned to be false (UNSAT, unsatisfiable), the number of gates in the circuit is incremented until the corresponding decision problem becomes true (SAT, satisfiable), that it, there is a solution for the decision problem. In this way, it is obvious that circuits with minimum numbers of gates are found. Also, other techniques such as, upper or lower bounds estimations can be used in the search process, which realize much more efficient methods.

However, in this searching method, the following problem can happen. Let us consider the reversible function shown in Fig. 2 and Table 2. As the implementations of this function, two networks consisting of Toffoli gates are shown in Fig. 3a, b. With exhaustive or intensive searches, we understand that there exist networks with 2 gates ($d = 2$, (a)) and 4 gates ($d = 4$, (b)), but there does not exist networks with 3 gates ($d = 3$). This example shows that even if a realization with d gates is found, its minimality cannot be confirmed by only checking that there is no implementation $d - 1$ gates. On the other hand, in order to prove that d gates is the minimum implementation, it is surely sufficient to show non-existence of networks with $d - 1$ and also $d - 2$ gates, as the following theorem holds (Theorem 1).

Theorem 1 *Let $f : B^n \rightarrow B^n$ be a reversible function from which a circuit should be generated. A Toffoli network consisting of d gates is a minimal implementation in terms of the number of gates, if no implementation with $d - 1$ gates or no implementation with $d - 2$ gates exist [6].*

The theorem says even if the decision problem is proven to be false (UNSAT) with d gates, still we need to check the decision problem with $d + 1$ gates, if the value of d is chosen in non-incremental ways.

Based on this, a pseudo code of an exact synthesis method based on the above discussion becomes the following. Using the above theorem, the values may not be

Fig. 3 Function which can be realized by networks with $d = 2$ or 4 gates but can not be realized with $d = 3$ gates

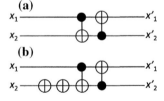

Table 2 An example truth table for outputs to inputs transformation

x_2	x_1	x'_2	x'_1
0	0	0	0
0	1	1	1
1	0	0	1
1	1	1	0
1	0	1	0

necessarily incremented by one. Instead it can jump up to say, $d1$ as long as it can be proven that both of the decision problems for $d1 - 1$ and $d1 - 2$ are shown to be UNSAT. But in the pseudo code, that is not used. Instead the number of gates is always incremented by one (line 12 in the algorithm).

Algorithm 1 An exact synthesis algorithm based on SAT solver

1: **procedure** EXACTSYNTHESIS($f : B^n \to B^n$)   ▷ f is given in the form of truth table
2:    $found = false$
3:    $d = 1$
4:    **while** $found \equiv false$ **do**
5:       $inst \leftarrow Encoding(f, d)$
6:       $res \leftarrow SATsolver(inst)$
7:       **if** $res \equiv SAT$ **then**   ▷ There is an implementation for f with d gates
8:          $A \leftarrow getSolution()$
9:          $generateNetworkFromSolution(A)$
10:          $found = true$
11:       **else**   ▷ There is no implementation for f with d gates
12:          $d \leftarrow d + 1;$
13:       **end if**
14:    **end while**
15: **end procedure**

The synthesis algorithm receives a truth table of the target reversible function f. It starts to try to implement f with just one gate, i.e., d is initialized to 1. If there is no implementation with d gates, d is incremented. This is repeated until an implementation is recognized. For each iteration, first the synthesis problem is represented as a Boolean satisfiability problem, and then it is solved by a SAT solver or a SMT solver.

If there exists a solution on the SAT/SMT problem, an implementation has been found, and it is generated from the solution as an instance of the general parametrized circuit. If there is no solution, it has been proven that there is no way to implement using d gates. By increasing d incrementally from d = 1, that is, minimality is guaranteed, when it starts with $d = 1$ and d is incremented one by one. Please note that due to Theorem 1 above, d should be incremented one by one. If d is incremented two by two, for example, even if the SAT problem becomes UNSAT for some value of d, there is a possibility that an implementation exists for $d - 1$. So the case for $d - 1$ must also be checked by SAT/SMT solvers.

Now the remaining issue is how to represent the decision problem in Boolean formulae or some theories which are decidable. For a reversible function, the synthesized circuit must satisfy three constraints. Satisfying these three constraints is equivalent to whether there exists a Toffoli gate network for the reversible function f using d gates. In order to describe these constraints, a set of variable vectors are introduced next.

First, for a Toffoli network consisting of k gates, the following variable vectors is defined. There are a number of lines in a Toffoli gate network, and there are multiple columns. These columns are essentially Toffoli gates and the computing in the network proceeds from the left to the right. That is, a righter column corresponds to a Toffoli gate having a larger number of levels. We call each location of the columns as depth. Here we assume that the target reversible function has n inputs and n outputs, and the maximum depth is d as there are d Toffoli gates.

Definition 8 Let $f : B^n \to B^n$ be a reversible function. The following variable vectors are defined:

1. $\mathbf{t^k} = (t^k_{\lceil log_2 n \rceil}, \ldots, t^k_1)$, with $0 \le k < d$, is a vector of Boolean variables showing a binary encoding of an integer $t^k \in \{0, \ldots, n-1\}$ which defines the selected target line for the Toffoli gate at depth k.

2. $\mathbf{c^k} = (c^k_{n-1}, c^k_{n-2}, \ldots, c^k_1)$, with $0 \le k < d$, is a vector of Boolean variables showing the control lines of the Toffoli gate at depth k. Assigning $c^k_i = 1$ with $(1 \le i \le n-1)$ means that line $(t^k + i) \bmod n$ becomes a control line of the Toffoli gate at depth k, as t^k is the target line.

Note that there are totally $n \cdot 2^{n-1}$ different Toffoli gates for a reversible function having n variables. This is the case since a Toffoli gate has exactly one target line and maximally $n - 1$ control lines. Therefore, there are in total n lines for placing a target line and 2^{n-1} combinations for control lines. Figure 4 shows 6 out of all $3 \cdot 2^{3-1} = 12$ possible Toffoli gates in a network with $n = 3$ variables. For each gate, the values of the vectors $\mathbf{t^k}$ and $\mathbf{c^k}$ are attached in the figure. For example, the values $\mathbf{t^k} = (01)$ and $\mathbf{c^k} = (01)$ indicate that line $[1]_2 = 1$ is the target line. Furthermore, because c_1 is assigned to 1, line $(1+1) \bmod 3 = 2$ becomes a control line. In contrast, because c_2 is assigned to 0, line $(1+2) \bmod 3 = 0$ does not become a control line.

Now, the variable vectors for the inputs and outputs of the network, as well as the internal signals, are defined.

Definition 9 Let $f : B^n \to B^n$ be a reversible function. Then, $\mathbf{x^k_i} = (x^k_{i_n}, x^k_{i_{(n-1)}}, \ldots, x^k_{i_1})$, with $0 \le i \le 2^n - 1$ and $0 \le k \le d$, is a vector of Boolean variables showing the inputs, internal gates, or outputs at depth k for row i of the truth table when implementing the target function f. The inputs and the outputs of the truth table for depth k correspond to the vectors $\mathbf{x^{k-1}_i}$ and $\mathbf{x^k_i}$, respectively.

Fig. 4 Examples of representation of Toffoli gates with $\mathbf{t^k}$ and $\mathbf{c^k}$

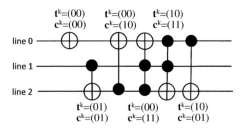

Fig. 5 Representation of the
decision problem

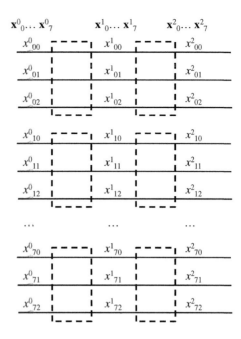

Figure 5 shows the variables in the above formulation for the constraints of a
function with $n = 3$ variables and depth $d = 2$. The positions for the Toffoli gates
to be synthesized should be located inside the dashed rectangles. For each of the
$2^3 = 8$ rows in the truth table, $n = 3$ lines in the network with the respective vectors
for inputs, internal gates, and outputs are used (i.e., overall $3 \cdot 8 = 24$ lines are considered). For each depth, all possible Toffoli gates can be defined with the respective
values for the vectors of variables, $\mathbf{t}^k$ and $\mathbf{c}^k$. For each depth, only one Toffoli gate
for each truth table row should be defined.

Based on these definitions on the constraints, the synthesis problem for a reversible
function f with d Toffoli gates can be formulated as follows:
*Is there an assignment for all variables of the vectors $\mathbf{t}^k$ and $\mathbf{c}^k$ such that, for each
line i, $\mathbf{x}_i^0$ is equivalent to the left side of the truth table, while $\mathbf{x}_i^d$ is equivalent to the
corresponding right side?*

The detailed conditions are shown below. The interpretation of these should be
referred to [6].

1. The input and output constraints set the input and output pair of each row of the
truth table given by the reversible function f

$$f(x) = \prod_{i=0}^{2^n-1} ([\mathbf{x}_i^0]_2 = i \wedge [\mathbf{x}_i^d]_2 = f(i))$$

2. The functional constraints for possible Toffoli gates which are selected by an
assignment to $\mathbf{t}^k$ and $\mathbf{c}^k$ are

$$f(x) = \prod_{i=0}^{2^n-1} \prod_{k=0}^{d-1} (\mathbf{x}_i^{k+1} = t(\mathbf{x}_i^k, \mathbf{t}^k, \mathbf{c}^k))$$

These conditions are saying that if, in line i at depth k, a Toffoli gate is used (i.e., $\mathbf{t}^k$ and $\mathbf{c}^k$ are completely assigned values), the variables in the next depth $k+1$ are computed from the variables at depth k with the Toffoli gate function $t(\mathbf{x}_i^k, \mathbf{t}^k, \mathbf{c}^k)$. The function $t(\mathbf{x}_i^k, \mathbf{t}^k, \mathbf{c}^k)$ is the function realized by a Toffoli gate with target line $t^k = [\mathbf{t}^k]_2$ and the control lines defined by $\mathbf{c}^k$.

3. The following conditions indicate that all illegal assignments to $\mathbf{t}^k$ must be excluded since not all values of $\mathbf{t}^k$ are necessary to enumerate all possible target lines

$$f(x) = \prod_{k=0}^{d-1} [\mathbf{t}^k]_2 < n$$

For example, for a network consisting of $n = 3$ lines, the target line is indicated with the two variables $\mathbf{t}^k = (t_2 t_1)$, as shown in Fig. 4. Then, the assignment $\mathbf{t}^k = (11)$ must be excluded from the solutions, since there is no line $[11]_2 = 3$.

The way to give these constraints are basically the same as LUT (Look Up Table) which is defined and used in the logic debugging discussed in the next chapter. LUT is a representation of a truth table of the target logic function using 2^n variables which show the values of rows in the truth table, and so the formulation is the same as above. By assigning values to these variables, all of the logic functions with n input (in total 2^{2^n} functions) can be represented. Here the goal is to identify the appropriate logic functions by SAT/SMT solvers as solutions on the assignments of values to these variables.

Based on these constraints, there can be different encoding possible when the Boolean decision problems by SAT/SMT are solved. For example, the conditions are simply transformed into CNF (Conjunctive Normal Form) for SAT solvers. Here all bit-vector variables and constraints must be converted into CNF using Boolean variables and clauses. It is known that this conversion to CNF take linear time and space with respect to the size of the original Boolean formula. The resulting CNF, however, is only a Boolean formula based on clauses, and due to the many auxiliary variables introduced during conversion, SAT solvers may not work so efficient.

Another encoding that avoids the conversion to the Boolean formulae can be considered, and the constraints are formulated as bit-vector logic inside SMT (Satisfiable Modulo Theory) solvers. All bit-vector variables and most of the operators are preserved as they are inside SMT solvers. Only a few additional variables are necessary when translating into the formula for SMT solvers. This can potentially significantly speed up the synthesis process.

With further careful consideration on the encoding as well as the effective use of SAT/SMT solvers, circuits having up to 10 gates or so can be optimally synthesized [6]. The method presented above has further been extended in [7]. Please see the paper for the details of the extensions.

5   Transformation Based Method

The transformation based method utilizes the fact that if a Toffoli gate is applied to the inputs or the outputs of a reversible function, the resulting functions always become reversible. So the synthesis problem can be recognized as finding a sequence of Toffoli gates which change a given reversible function to the identity function. Since a sequence of gates can be applied either to the inputs or the outputs, the synthesis process can proceed from outputs to inputs, inputs to outputs or, in both directions in a combined way [8].

So a basic and naive algorithm would be to synthesize the circuit in one direction, specifically from outputs to inputs, and it is given below first. This algorithm is guaranteed to finish without introducing unnecessary garbage outputs, and the synthesized circuits have at most $n2^n$ gates.

Transformation Based Synthesis: Basic Algorithm

1. Step 1:
   If $f(0) \neq 0$, invert the outputs which generate 1 for $f(0)$. A complement operation requires one gate. The function after this, f^+ satisfies $f^+(0) = 0$.
2. Step 2:
   Consider each i in turn for $1 \leq i < 2^n - 1$, that is, for all possible input values for the function. Let f^+ denote the reversible specification currently being considered. If $f^+(i) = i$, no transformation is required as it keeps the value and it satisfies the requirement to be a reversible function. So no Toffoli gate is implemented on i. Otherwise, gates are required to transform the specification to a new specification which satisfies $f^{++}(i) = i$. The required gates should transform as follows: $f^+(i) \rightarrow i$.
   Let p be the bit string having 1's in all the positions where the binary expansion of i is 1 whereas the expansion of $f^+(i)$ is 0. These are the bits which have value 1 and must be added in transforming $f^+(i) \rightarrow i$. Conversely, let q be the bit string having 1's in all the positions where the expansion of i is 0 whereas the expansion of $f^+(i)$ is 1. q identifies the bits to be removed through some operation.
   For each j where $p_j = 1$, use a Toffoli gate with the control lines which correspond to all the outputs in the positions where the expansion of i is 1 and the target line which corresponds to the output in position j. Also, for each $q_k = 1$, use a Toffoli gate with the control lines which correspond to all the outputs in the positions where the expansion of $f^+(i)$ is 1 and the target line which corresponds to the output in position k.

For each i in $1 \le i < 2^n - 1$, The transformation in Step 2 performs: $f^+(i) \to i$ using the above mentioned sequence of Toffoli gates. i is considered in order, and Step 1 handles the case when $i = 0$, and therefore, $f^+(j) = j$, $0 \le j < i$ is satisfied. Please be reminded that none of the Toffoli gates used in Step 2 influence the equality: $f^+(j) = j$, $0 \le j < i$. That is, once a row of the truth table for the specification is converted into the new value, it remains at that value regardless of the conversions required for later rows.

As it may not be easy to intuitively understand the algorithm above, let us discuss through an example. An illustrative example of a transformation based approach is shown in Table 3. Columns A in the table are the given specification. As the process of Step 1, x_1 is applied the application of TOF_1 gate ($TOF_1(x_0)$) which generates in columns B in the table. Underlined values are the results of the complement of x_1 At this stage the required condition for all of $f^+(i)$, $0 \le i \le 4$ is already satisfied, and so no further action on these is required. Then $TOF_3(x_3^1, x_2^1, x_1^1)$ is applied in order to partially realize $f^+(5) \to 5$, that is, changing the rightmost position to 1. This generates in columns C in the table. In order to convert the center 1 to 0, $TOF_3(x_3^2, x_1^2, x_2^2)$ is used. As the last stage, in order to realize $f^+(6) \to 6$, $TOF_3(x_3^3, x_2^3, x_1^3)$ is used. Please be reminded that the generated gates are identified in order from the output side to the input side. Figure 6 shows the final circuit. This is a basic algorithm and is rather straightforward. It can be implemented easily, but the complexity of this algorithm is $n2^n$. We can easily understand that the algorithm always terminates successfully and generates a circuit for the given specification.

A sequence of transformations (gate generation) are applied from the outputs to the inputs in the above example. For a reversible function as specification, the inverse way of processing can be considered. That is, a sequence of transformations are applied from the inputs to the outputs. Then the better way can be used. Also, the

Table 3 An example truth table for outputs to inputs transformation

			A			B			C			D			E		
x_3	x_2	x_1	x_3^0	x_2^0	x_1^0	x_3^1	x_2^1	x_1^1	x_3^2	x_2^2	x_1^2	x_3^3	x_2^3	x_1^3	x_3^4	x_2^4	x_1^4
0	0	0	0	0	1	0	0	$\underline{0}$	0	0	0	0	0	0	0	0	0
0	0	1	0	0	0	0	0	$\underline{1}$	0	0	1	0	0	1	0	0	1
0	1	0	0	1	1	0	1	$\underline{0}$	0	1	0	0	1	0	0	1	0
0	1	1	0	1	0	0	1	$\underline{1}$	0	1	1	0	1	1	0	1	1
1	0	0	1	0	1	1	0	$\underline{0}$	1	0	0	1	0	0	1	0	0
1	0	1	1	1	1	1	1	$\underline{0}$	1	1	1	1	0	1	1	0	1
1	1	0	1	0	0	1	0	$\underline{1}$	1	0	1	1	1	1	1	1	0
1	1	1	1	1	0	1	1	$\underline{1}$	1	1	0	1	1	0	1	1	1

Fig. 6 The synthesized circuit from Table 3

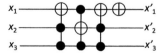

Table 4 An example truth table for bidirectional transformation

			A			B			C			D		
x_3	x_2	x_1	x_3^0	x_2^0	x_1^0	x_3^1	x_2^1	x_1^1	x_3^2	x_2^2	x_1^2	x_3^3	x_2^3	x_1^3
0	0	0	1	1	1	0	0	0	0	0	0	0	0	0
0	0	1	0	0	0	1	1	1	0	0	1	0	0	1
0	1	0	0	0	1	0	1	0	0	1	0	0	1	0
0	1	1	0	1	0	0	0	1	1	1	1	0	1	1
1	0	0	0	1	1	1	0	0	1	0	0	1	0	0
1	0	1	1	0	0	0	1	1	1	0	1	1	0	1
1	1	0	1	0	1	1	1	0	1	1	0	1	1	0
1	1	1	1	1	0	1	0	1	0	1	1	1	1	1

two ways can be combined, that is, some gates are used as transformations from the inputs to the outputs, while other gates are used as transformations from the outputs to the inputs. This combined one is called bidirectional transformations.

An example of bidirectional transformations is shown in Table 4. Columns A in the table is the target reversible function. The basic algorithm with transformations from outputs to inputs discussed above would require that all of x_3, x_2, x_1 are complemented in order to establish $f^+(0) = 0$. On the other hand, there is an alternative way which complements x_1 only, i.e., uses the gate $TOF_1(x_1^0)$ to the input side. By using this gate, and then by reordering the specification in such a way that the input side is again in standard truth-table order, the specification in columns B in the table is obtained. In the case of starting with the output side, the mapping $f^+(i) = 7 \rightarrow 1$ should be established. However, from the input side we only have to exchange rows 1 and 3, and it can be implemented by using the gate $TOF_2(x_1^1, x_2^1)$. By doing that operation and by again reordering the input side into standard order, columns C in the table are obtained. At this stage, the transformation from the output side and the transformation from the input side generate the same gate $TOF_3(x_1^2, x_2^2, x_3^2)$, columns D in the table show the resulting target functions.

The application result of this bidirectional approach is the circuit shown in Fig. 7a. It has 3 gates. On the other hand, the application result of the transformations from the outputs to the inputs is the circuit shown in Fig. 7b. It has 5 gates.

In general, when $f^+(i) \neq i$, there are two ways to proceed:

1. Use Toffoli gates to the outputs to satisfy $f^+(i) \rightarrow i$
2. Use Toffoli gates to the inputs to satisfy $j \rightarrow i$ where j is chosen such that $f^+(j) = i$. Since the synthesis process proceeds with i in order, such $j(> i)$ must always exist.

Appropriate choices of the above (1) and (2) should be enforced in order to generate smaller circuits.

There may be redundancy in the circuits generated by the methods mentioned above. In the examples of the above, the three serially connected gates, $TOF_2(x_2, x_1)$,

Fig. 7 The synthesized
circuit from Table 4

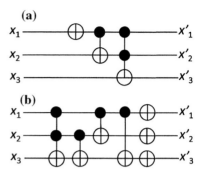

$TOF_1(x_2)$, $TOF_1(x_1)$ can be replaced by the two serially connected gates, $TOF_1(x_2)$, $TOF_2(x_2, x_1)$, as they are doing the same computations as the whole. This is an equivalent preserving rule which can be utilized for optimization and is called "template" in [8]. Similar methods have been used in conventional logic synthesis and are called rule-based logic optimization. Rule-based logic optimization may use large numbers of rules, such as over 200 rules, depending on the target technologies.

A template consists of a serially connected gates to be matched and the serially connected gates to be substituted when a match is found.

The lines in the template are to be matched with real lines in the circuit satisfying the logical consistency on the matching. There may be several templates which can match the real lines in the target circuit. This is realized by first mapping the widest target template gate with a gate in the circuit and then looking for the other target gates in the circuit using the line mapping derived from the widest gate. As the order of the control lines to a Toffoli gate is irrelevant when matching, there are $c!$ line mappings which must be considered where c is the number of control lines for the widest gate.

There are similar problems in rule-based logic optimization and also technology mapping processes in conventional logic synthesis methods. There in order to identify the matched rules and gates/cells in the cell libraries of the target semiconductor technology. Because of symmetry in inputs for various gates/cells, such as a 12 inputs AND gate, basically all permutation of the 12 inputs must be tried. If we check one by one, this takes a very long time. There are basically two ways to go. One is based on only structural analysis and the set of matchable gates/cells are generated for all permutations of the inputs of the gate/cell. Although this works very efficient (as it is purely based on structural matching), the numbers of tries for matching may become very large. The other way is to check the matching based on logical analysis such as SAT checking. Although this needs only one try for matching for each gate/cell, the Boolean reasoning process may not be so quick in general depending on the type of functions in the circuit and gates/cells. Same approaches can be applied also in the case of template-based optimizations.

Besides the matching problem, we need to look for the target gates to be matched. This includes the initial match to the widest gate out of all gates in the entire circuit.

If all target gates are found, they are checked to be replaced so that they are adjacent matching the template either in the forward or reverse direction. If this checking is positive, the matched gates are replaced with the new gates specified by the template. There can be a reverse match, and the new gates are substituted in reverse order.

When traversing the target gates, the matching procedure takes account of **Theorem 2** which follows directly from the definition of control lines in Toffoli gates. If two gates can be swapped, the swapped one is also checked to be matched with templates.

Theorem 2 *Two gates $TOF_k(x_1, x_2, \ldots, X_{k-1}, x_k)$ and $TOF_l(y_1, y_2, \ldots, y_{l-1}, y_l)$ adjacent in a circuit can be swapped, if and only if $x_k \notin \{y_1, y_2, \ldots, y_{l-1}\}$ and $y_l \notin \{x_1, x_2, \ldots, x_{k-1}\}$.*

The matching procedure basically tries all target gates for each template. When a template match is found, the substitution is performed based on the matched template. Then the matching process restarts, since a substitution may make a template which is rejected earlier be applicable.

There are different classes of templates that can be defined.

1. Two inputs involving a swap operation
2. Two input gate reductions without any swap operation
3. Based on transformation rule
4. Symmetric templates
5. Controlled SWAP

Examples of templates in the above classes are shown in Figs. 8 and 9. By using these templates as rules of transformations, reversible circuits can potentially be reduced significantly, although the final quality, such as the size of reversible circuits largely depends on the order of applications of the templates. Same as the case of rule-based logic optimization in the conventional logic synthesis, it is not straightforward to see the order of applications of various rules or templates, and the initial circuits tends to influence the lots of the final circuits.

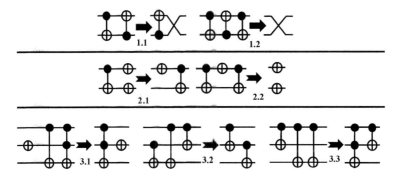

Fig. 8 Examples of templates in classes 1–3

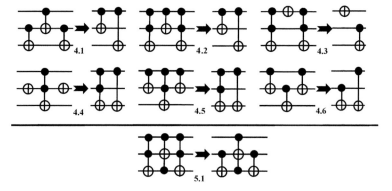

Fig. 9 Examples of templates in classes 4–5

On the other hand, again as is the case of conventional logic synthesis processes, rule-based logic optimization can further optimize circuits which are synthesized by other techniques and is actually used intensively in the commercialized logic synthesis tools. In that sense, template based synthesis can plan an important role in reversible logic synthesis.

6 Decision Graph Based Method

It is well known that Boolean functions can be efficiently and compactly represented by Binary Decision Diagrams [4] for many useful logic functions which are commonly used in hardware and also software developments. The only exception is the multiplication function where BDD must explode exponentially with respect to the number of inputs. Based on the experiences with BDD, all the other functions useful for human designs are said to be very compact if good variable ordering are used. Also thanks to the 64-bit processors, large BDD, such as the one having over 10,000,000 nodes can be manipulated and analyzed, as BDD is a canonical representation for logic functions.

One remark here is that although the size if BDD for the functional practically appearing in hardware and software developments a reasonably compact and can be efficiently manipulated on the computer, the number of BDD nodes can be much larger than the number of gates for conventional multi-level logic circuits. As conventional multi-level logic circuits represent arbitrary logic expressions, they are not at all canonical, but they can be much mode compact.

Having a BDD, $G = (V, E)$ (where V denotes the set of vertices and E denotes the set of edges in the graph), a reversible circuit can be generated by traversing the BDD and substituting each node $v \in V$ with a cascade of reversible gates [9]. The BDD for exclusive-OR operation is shown in Fig. 10a. Its circuit implementation by

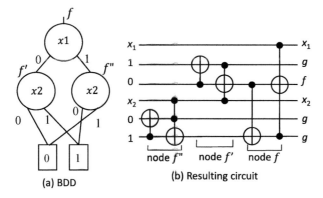

(a) BDD

(b) Resulting circuit

Fig. 10 BDD for full adder and its Toffoli gate implementation

Fig. 11 General case of
BDD node replacement

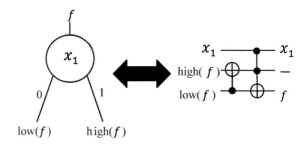

Toffoli gates is shown in Fig. 10b. The type of the node v determines how to cascade the gates.

The general case of a node in a BDD is shown in Fig. 11, and its implementation with two Toffoli gates is also shown in the figure. Based on the definition of Toffoli gates, one can confirm that the logic function realized at f in the right side is equal to $low(f) \cdot \overline{x_1} + high(f) \cdot x_1$. Following this substitution, a complete Toffoli network can be generated from a BDD if the BDD has no sharing of nodes. When generating an implementation, inputs have to be set to constants, if the respective $low(v)$ or $high(v)$ edge goes to a terminal node (either 0 or 1). If the BDD is a kind of a tree instead of a graph, this can be simply applied to generate a reversible circuits with Toffoli gates. Please note that BDD representations for logic functions are compact because of the sharing of sub-graphs in a BDD, and so generally speaking a BDD has a number of sharing sub-graphs.

Let us take the BDD shown in Fig. 10a. Please note that when the conversions given in Fig. 11 are applied to each node of the BDD, Fig. 10b becomes the resulting Toffoli gate based circuit. The BDD is actually sharing the sub-graph and nodes by which its size becomes much more compact. If the conversions are applied to the BDD having shared nodes, the resulting circuits are not correct at the points of the fan-outs in the circuits.

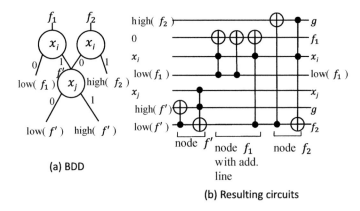

Fig. 12 Toffoli gate network example for a BDD with a shared node

The values on the signals in the circuit corresponding to the shared nodes must be preserved until all of the fan-out are converted. The same must be applied to the signals corresponding to the case-split variables of the nodes, as they are often required more than once.

There are two edges going to the node of variable x_1 in Fig. 12a. That is that node is shared by the two other nodes. Therefore, the values of that node must be kept until it is used twice by the other two nodes. An additional line and an adjusted cascade of gates are inserted into the Toffoli network, in order to implement this as a reversible circuit, since the value of node f' is used twice (by nodes f_1 and f_2), an additional line (the second line in Fig. 12b) and the cascade of the Toffoli gates are used to substitute node f_1. In doing so, the value of f is still available so that the substitution of node f_2 can be applied. The resulting circuit is given in Fig. 12b.

Now let us discuss which cascade of the Toffoli gates should be used when a node is referred multiple times from other nods. See the example BDD in Fig. 13a. The node which corresponds to the function f' is referred two time as shown in the figure. The Toffoli network shown in the figure is the cascade of gates to be used when a node is shared. The logical relationship from the Toffoli network in the figure becomes the followings:

$t_{11} = ((x_i \wedge low(f_1)) \to \overline{0}) \wedge (\overline{x_i \wedge low(f_1)} \to 0)$

$t_{12} = x_i$

$t_{13} = low(f_1)$

$t_{21} = ((t_{13} \to \overline{t_{11}}) \wedge (\overline{t_{13}} \to t_{11})$

$t_{22} = t_{12}$

$t_{23} = t_{13}$

$t_{31} = ((t_{23} \wedge f') \to \overline{t_{21}}) \wedge (\overline{t_{23} \wedge f'} \to t_{21})$

$t_{32} = t_{22}$

$t_{33} = t_{23}$

By substituting and simplifying the above equations, one can make sure they establish the relationship derived from the BDD structure.

Fig. 13 A BDD having a shared node

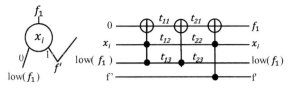

(a) BDD   (b) Resulting circuits

Table 5 Part of the truth table for node v with $high(v) = 0$ without additional line

x_i	$low(x_i)$	f	–
0	0	0	0
0	1	1	1
1	0	0	1
1	1	0	(1)

Table 6 Part of the truth table for node v with $high(v) = 0$ with an additional line

0	x_i	$low(x_i)$	f	x_i	$low(x_i)$
0	0	0	0	0	0
0	0	1	1	0	1
0	1	0	0	1	0
0	1	1	0	1	1

Please note that the three Toffoli gates are used instead of two gates for the case of non-sharing. Also there is one additional line (the bottom one) to keep the function for f'. This cascade of the three Toffoli gates can be used each time a node is shared by multiple nodes in the BDD.

Using these substitutions or conversions a given BDD can be transformed to a Toffoli gate based circuit. However, there are better and more efficient ways for substitutions, if the successor of a node in a BDD is a terminal node (constant 1 or 0).

For example, a node v with $low(v) = 0$ (fourth row) can be synthesized with only one Toffoli gate as shown in Fig. 15d. Various situations when one of the child nodes is a terminal (constant 1 or 0) are shown in Fig. 15. They include identity node shown in F of the Figure, i.e., a node with $low(v) = 0$, $high(v) = 1$, and select a variable x_i. Furthermore, due to the additional line, which has to be added if either $low(v)$ or $high(v)$ is a terminal, it is also possible to preserve all inputs of a node. In particular for shared nodes, this allows better substitutions. Also, please note that BDD having negative edge generates smaller circuits in general, and the rules for them can be similarly defined.

Let us discuss on a case of Fig. 15. The truth table for the BDD with $high(v) = 0$ is shown in Table 5. The first three rows of the truth table follow the condition for a reversible function as shown in the figure. But the last row does not follow, since f is 0 in the last row (there are three 0s for the value of f). Therefore, an additional line is required to realize a reversible circuit as shown in Table 6.

The above discussions can be summarized as the following synthesis algorithm. First, a BDD for the target function f from which a reversible circuit is to be synthesized is constructed. This can be efficiently implemented with the state-of-the-art BDD packages. Secondly, the constructed BDD, $G = (V, E)$ is traversed from the roots in a depth-first way. For each node $v \in V$, there are four cases:

1. When a node v represents the identity of a primary input (i.e., it is a select input): The circuit does not need any cascade of gates, since the identify function is simply the same as the input itself.
2. When a node v contains at least one edge ($low(v)$ or $high(v)$) leading to a terminal node (constant 1 or 0):
   The circuit needs the substitutions shown in Fig. 15, which uses smaller numbers of gate than the general substitutions.
3. When the successors of a node v (i.e., $low(v)$ and $high(v)$) are shared nodes and are still needed:
   The circuit needs the substitution which preserves the values of all input signals.
4. When none of the above cases hold:
   The circuit needs a cascade of gates shown in Fig. 15a.

Now the above synthesis method is applied to an example shown in Fig. 14. $f_1 = x_1 \wedge x_2$ and $f_2 = x_1 \vee x_2$ are the specifications. First, the above synthesis method reaches node f'. Since f' represents the identity function with respect to x_2, no gates is added. The third line of the circuit in Fig. 14b is used for storing both of the value of the primary input x_2 and the value of f'. Then, for node f_1, the substitution shown in Fig. 15d is applied. This not only reduces the number of gates.but also preserves the value of f' which is still needed to create a circuit for f_2. Finally the node f_2 can be substituted, and the synthesis finishes.

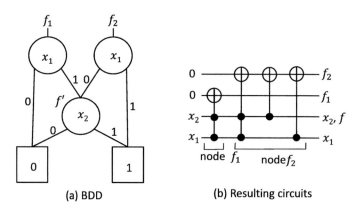

(a) BDD   (b) Resulting circuits

Fig. 14 Toffoli gate network example for a BDD with a shared node

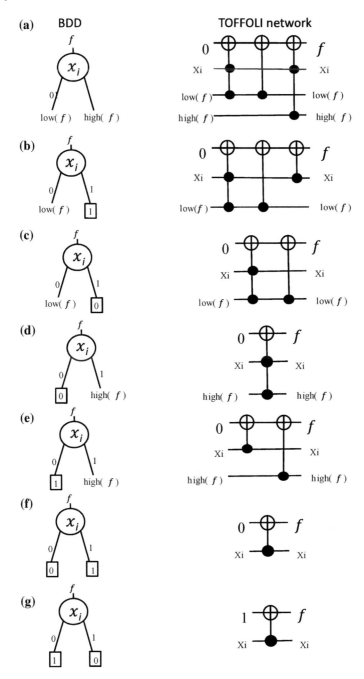

Fig. 15 Toffoli gate network example for a BDD with a shared node

Since, each node of the BDD is only substituted by a cascade of gates, the proposed method has a linear worst case run-time and linear memory complexity with respect to the number of nodes in the BDD. Unfortunately, the numbers of nodes in the BDD for a given logic function is larger than the numbers of gates for the corresponding multi-level circuits using the traditional technologies such as CMOS. Therefore, although this BDD based synthesis method can work for large circuits, the synthesized reversible circuits may not be compactly represented.

In the above BDD based synthesis method, additional lines may have to be introduced. For the reduction of the numbers of such additional lines, an improved method is also proposed [10].

7   LUT Based Method and Hierarchical Approach

One big problem on the methods discussed so far is how to synthesize large circuits without introducing much extra (or redundant) lines or gates. BDD based synthesis can work for relatively large functions, but the quality of synthesized circuits may not be always good, as BDD generally needs larger numbers of nodes than the numbers of gates in conventional multi-level logic synthesis. All the other methods above cannot basically deal with large circuits at all.

On the other hand, conventional logic synthesis tools, such as ABC [3] processes large circuits by introducing "windowing" or partitioning lager circuits into smaller ones which are actually manipulated by the logic synthesis tools. At a time, the logic synthesis tools are concentrating only on one particular portion (windows or partition) of the entire circuit, and the optimized results are just combined together in order to obtain the final circuit. Such divide-and-conquer approach is the key for the logic synthesis of large and practical sizes of circuits.

In the field of logic synthesis for reversible functions, such hierarchical approach is introduced in [11]. It has been shown that hierarchical reversible logic synthesis methods based on logic network representations are able to synthesize large arithmetic designs. The underlying idea is to map sub-networks into reversible networks. If the sub-networks are small enough, one can use the reversible synthesis methods discussed so far which is not so scalable including the ones based on Boolean satisfiability.

However, logic networks differ significantly from reversible logic networks in terms of their structures, such as extra lines appearing quite often in the above. This introduces a problem for simple hierarchical synthesis methods. For example, the outputs of the reversible circuits in quantum computers should be either a primary input value, a primary output value, or a constant, and they cannot expose an intermediate result to an output line, which is referred to as a garbage output. State-of-the-art algorithms such as the approach shown in [11] do not explicitly consider techniques on the reduction of garbage outputs. Moreover, in order to use the circuit in a quantum computer, one needs to apply a technique called "Bennett trick" [2], where the number of gates must be doubled and an additional line must be added for each primary output.

Before discussing the hierarchal synthesis method, single-target gate is introduced. A reversible logic circuit implements a reversible function, and reversible circuits are a cascade of a number of reversible gates. The most general gate which can be considered is the single-target gate. A single-target gate $T_c(\{x_1, \ldots, x_k\}, x_{k+1})$ has control lines $x_1, \ldots, x_k$, a target line x_{k+1}, and a control function $c : B^k \rightarrow B$. It implements the reversible function $f : B^{k+1} \rightarrow B^{k+1}$ with $f : x_i \rightarrow x_i$ for $i \leq k$ and $f : x_{k+1} \rightarrow x_{k+1} \oplus c(x_1, \ldots, x_k)$. This definition is conceptually similar to a Toffoli gate. All reversible functions can be realized by cascades of single-target gates just like cascades of Toffoli gates.

Now a hierarchical synthesis approach based on k-feasible Boolean logic network is introduced. k-feasible network is a circuit consisting of gates whose numbers of inputs are k or less. That is, it is a logic network in which every gate has at most k inputs. These are the same as k-LUT (Look Up Table) which can represent all possible truth tables up to k inputs. LUT is a commonly used basic gate for Field Programmable Gate Arrays (FPGA).

If a given network does not satisfy the requirement for k-feasible network, circuit restructuring can be performed to let it be k-feasible. There are various ways of such circuit restructuring implemented on conventional logic synthesis tools including ABC from University of California Berkeley [3]. Figure 16 shows an example of such restructuring process assuming that $k = 4$. The upper circuit is transformed into k-feasible circuit in the bottom. The inside matrices with 0, 1, and - denote the functions to be realized by the LUTs in sum-of-products (SOP) formats.

Fig. 16 Conversion of a given circuit into k-feasible circuit

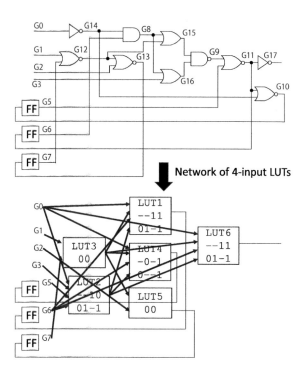

Normally each gate is better to have as many as inputs not exceeding k. The upper circuit in the figure is also a k-feasible circuit, since all gates have two inputs or less. In order to reduce the total number of gates in the circuit, however, each gate is better to have a larger number of inputs in general. Also, some of the original gates may be duplicated in the translation into the k-feasible circuit. So the translation may not be so straightforward in order to keep the numbers of LUTs as small as possible after translation.

There is a useful property that a one-to-one relationship between a k-input LUT in a logic network (k-feasible network) and a reversible single-target gate with k control lines in a reversible network exists. A single-target gate has a control function and a single target line which is inverted if and only if the control function evaluates to 1 as defined above.

The hierarchical synthesis method first generates a skeleton for the target reversible circuit. It consists only of single-target gates, and the number of required additional lines is already fixed and final. This means there is no garbage outputs. Then, each single-target gate is converted into a corresponding reversible circuit using any algorithm, which could be anyone discussed above or shown in the literature [12–16]. It is straightforward to parallelize and shorten the time for this second step as the synthesis problems are independent with one another. This two-step algorithm can be applied to find reversible circuits for several floating point arithmetic networks up to double precision computing [17]. Floating point arithmetic circuits are generally fairly large ones even in the state-of-the-art microprocessor circuits, occupying 20% or more area in the chips. This means the hierarchical method discussed here can even be applied to the real-life designs. From the synthesized circuits the cost for their use in, for example, quantum computers can be accurately estimated. Most of the existing synthesis methods do not give such information, especially for arithmetic circuits.

The details of the above hierarchical synthesis method are given below by illustrating the general idea on how to map LUT networks into reversible circuits. For this purpose, the LUT network shown in Fig. 17 is used as an example.

The example network has 5 primary inputs $x_1, \ldots, x_5$, 5 LUTs with names LUT1 to LUT5, and 2 primary outputs, y_1 and y_2. The two outputs are generated from LUT3 and LUT5, respectively. A straightforward way to translate a LUT into a reversible circuit is to use one single-target gate for each LUT in topological order of the circuit.

Given a LUT network of Fig. 17, the skeleton for its reversible circuit becomes the one shown in Fig. 18. The correspondence between Figs. 17 and 18 are in the following way. As LUT1 has x_1 and x_2 as inputs, there is a corresponding box in Fig. 18 which is indexed as 1. This box has all the control lines, depending on their values, the value of the additional line is controlled. So the box has a vertical line to that additional line as shown in the figure. The additional line's initial value is set to 0. In the same way, for LUT2, there is a box indexed as 2 and its additional line in Fig. 18.

For LUT3, there is a corresponding box whose inputs are x_1 and the output of LUT1 which is the additional line associated with the box 1. This process continues until the boxes for all LUT, LUT1-LUT5, are created with associated additional lines

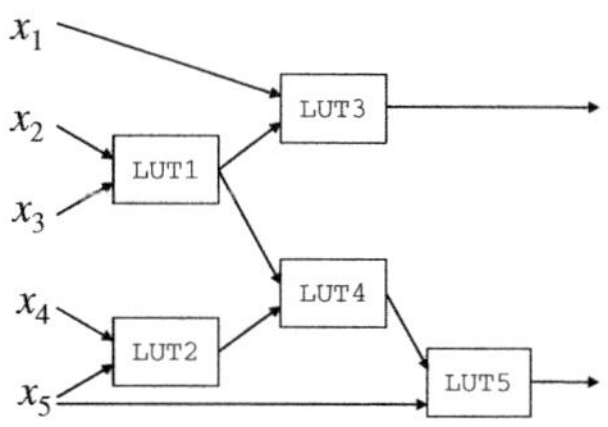

Fig. 17 An example of
2-feasible circuit

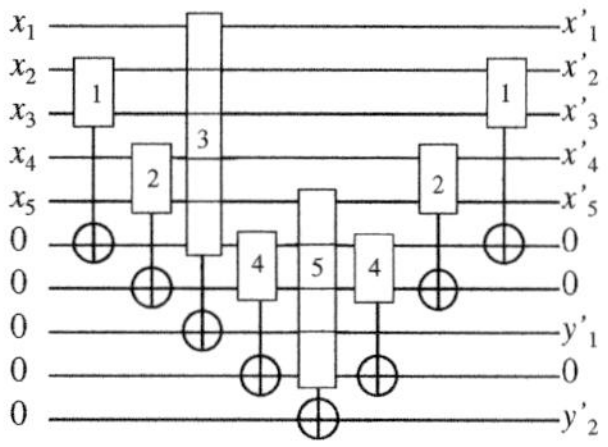

Fig. 18 Toffoli network
generated from the circuit in
Fig. 17 with the order of 1, 2,
3, 4, 5

as shown in Fig. 18. During that process, the initial values for the additional lines are all set to 0. Please note that in the figure, single-target gates are used rather than Toffoli gates. With these five gates, the primary outputs y_1 and y_2 are realized at line 8 and 10 of the reversible circuit.

Unfortunately after processing the five gates, the reversible circuit has garbage outputs (not related to the real outputs) on lines 6, 7, and 9 shown in Fig. 18. These perform the functions which are used in the LUTs inside the circuit. It is definitely better to have fewer numbers of such garbage outputs. This is because the result of the calculation is enlarged by the intermediate results, and so they cannot be discarded and recycled without decreasing the quality of the circuits [18]. A kind of uncomputation) on the intermediate results should be performed if ever possible, by re-applying the single-target gates for the LUTs in reverse topological order. In Fig. 18 the last 3 gates work for the uncomputations of the intermediate results at lines 6, 7, and 9. Based on this observation we derive the following theorem.

Theorem 3 *When realizing a LUT network with d gates by a reversible circuit that uses single-target gates for each LUT, at most d additional lines are required.*

There is, however, a better approach. Once the computation for a primary output is finished, uncomputations are applied to the LUTs that are not used any longer by other outputs. The lines which are applied uncomputations restore 0 values that can be used instead of creating an additional line. In the case of the 2-feasible circuit in Fig. 17 first primary output y_2 is computed. After that, the uncomputations for LUT4 and LUT2 are applied, as they are not in the logic cone of primary output y_1. The freed additional line can be used for the single-target gate realizing LUT3. This observation leads to a theorem providing a lower bound.

Theorem 4 *Given a LUT network with m outputs, let l be the maximum cone size over all outputs. When realizing the LUT network by a reversible circuit that uses single-target gates for each LUT, at least l additional lines are required.*

Fig. 19 Toffoli network
generated from the circuit in
Fig. 17 with the order of 1, 2,
4, 5, 3

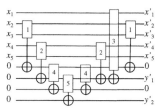

That is, we start by synthesizing a circuit for the output with the largest cone. Let us assume that this cone contains l LUTs. That can be synthesized using l single-target gates. From these l gates, $(l-1)$ gates can be applied the uncomputations (all except the LUT computing the primary output). Therefore, it restores $(l-1)$ lines which hold constant 0 values. It is easily understood that the exact number of required lines may be a bit larger, since all output values need to be kept. Further, it is better to make use of logic sharing and use at most two single-target gates for each LUT in the network. As can be seen from the previous discussion, the number of additional lines roughly corresponds to the number of LUTs. Hence, we are interested in logic conventional synthesis algorithms that minimize the number of LUTs, which have already been explored by many researchers under the context of logic synthesis for LUT networks. Those may be used for further optimization of the method discussed above.

If this method is applied to the circuit in Fig. 17, the order of processing LUTs becomes 1, 2, 4, 5, 3 instead of the original order 1, 2, 3, 4, 5. This results in the reversible circuit shown in Fig. 19 which has one less additional lines compared to the reversible circuit generated in Fig. 18.

The discussed method works well for large functions including the functions for floating point computations, which occupies 20% or more area even in the modern microprocessors. Although the method can be further improved by considering sizes of LUT and others, it is basically becoming very practical in terms of the sizes of circuits that can be synthesized. It may be applied to program Quantum computers with large numbers of Qbits.

8   Conclusions

In this chapter, we have discussed varieties of logic synthesis methods including, exact, transformation, BDD-based, and hierarchical approaches. Although each one has advantages and disadvantages, basically methods except for hierarchical approach are applicable only fairly small circuits.

On the other hand, the hierarchical approach gives a way to partition or windowing a large circuit, and after that, any methods discussed in this chapter can basically be applied. This is essentially the same approach used in conventional logic synthesis tools, such as ABC. With this combination of the approaches, reversible circuit synthesis for large functions can become practical. The experimental results by the various logic synthesis methods would be said to be applied to only toy examples with the exception of the hierarchical approach applied to real floating point calculation circuits which are large and practical. It is expected that more and more practical evaluation will be made for logic synthesis methods for reversible functions.

Also, logic debugging methods targeting reversible circuits are briefly discussed. As the reversible circuits become more popular, debugging and ECO would become more critical, and more intensive researches are expected.

References

1. Landauer R (1961) Irreversibility and heat generation in the computing process. IBM J Res Dev 5(3):183–191
2. Bennett C (1973) Logical reversibility of computation. IBM J Res Dev 17(6):525–532
3. Brayton RK, Mishchenko A (2010) ABC: an academic industrial-strength verification tool. In: 22nd international conference on computer aided verification (CAV 2010), pp 24–40
4. Bryant R (1986) Graph-based algorithms for boolean function manipulation. IEEE Trans Comput C-35(8)
5. Shende VV, Prasad AK, Markov IL, Hayes JP (2002) Reversible logic circuit synthesis. In: ACM/IEEE international conference on computer aided design (ICCAD), San Jose, USA, pp 125–132
6. Große D, Wille R, Dueck GW, Drechsler R (2009) Exact multiple control Toffoli network synthesis with SAT techniques. IEEE Trans CAD 28(5):703–715
7. Wille R, Soeken M, Przigoda N, Drechsler R (2012) Exact Synthesis of Toffoli Gate Circuits. In: IEEE 42nd international symposium on multiple-valued logic. Victoria, Canada, pp 69–74 (with Negative Control Lines)
8. Miller DM, Maslov D, Dueck GW A (2003) Transformation based algorithm for reversible logic synthesis. In: ACM/IEEE design automation conference. Anaheim, USA, pp 318–323
9. Wille R, Drechsler R (2009) BDD-based synthesis of reversible logic for large functions. In: ACM/IEEE design automation conference. USA, San Francisco pp 270–275
10. Soeken M, Taguea L, Dueckc GW (2016) RolfDrechsler: ancilla-free synthesis of large reversible functions using binary decision diagrams. J Symb Comput 73:1–26
11. Soeken M, Chattopadhyay A (2016) Unlocking efficiency and scalability of reversible logic synthesis using conventional logic synthesis. In: ACM/IEEE design automation conference
12. Iwama K, Kambayashi Y, Yamashita S (2002) Transformation rules for designing CNOT-based quantum circuits. ACM/In: IEEE design automation conference. New Orleans, Louisiana, USA
13. Kerntopf P (2004) A new heuristic algorithm for reversible logic synthesis. In: ACM/IEEE design automation conference
14. Maslov D, Dueck GW, Miller DM (2005) Toffoli network synthesis with templates. IEEE Trans CAD 24(6):807–817
15. Gupta P, Agrawal A, Jha NK (2006) An algorithm for synthesis of reversible logic circuits. IEEE Trans CAD 25(11):2317–2330
16. Shende VV, Prasad AK, Markov IL, Hayes JP (2003) Synthesis of reversible logic circuits. IEEE Trans CAD 22(6):710–722
17. Soeken M, Roetteler M, Wiebe N, De Micheli G (2017) Hierarchical reversible logic synthesis using LUTs. In: ACM/IEEE design automation confernece. Austin, USA
18. Nielsen MA, Chuang IL (2010) Quantum computation and quantum information. Cambridge University Press, Cambridge

Search-Based Reversible Logic Synthesis Using Mixed-Polarity Gates

S. C. Chua and A. K. Singh

Abstract Synthesis methods of irreversible circuits cannot be used for reversible circuits because of their logical differences. An algorithm for the synthesis of reversible circuits using its Positive-Polarity Reed–Muller (PPRM) expansions is presented in this chapter. The proposed algorithm used Hamming Distance (HD) approach to select the transformation path. A variety of reversible gates are selected through finding the possible matching reversible gate for path selection. It has the capability to allow the algorithm to synthesize reversible function in terms of quantum cost, which is a challenging task for existing synthesis algorithms. The proposed algorithm has been applied to synthesize all 3-variable reversible functions and has shown to obtain a good result. From the experimental results, it has been shown that with the m-NCT gate library added, the results are improved significantly.

1 Introduction

In the past few years, synthesis of reversible logic has been rigorously studied in the research centered on the synthesis of digital circuit by applying NCT library gates. NCT library gates have been widely studied and explored for several years. The existing Toffoli gate has been extended by increasing the size of variable for observing more possible synthesis outcome. Recently, the researcher has started to insert polarity control capability in Toffoli gate operations. Mixed- polarity Toffoli gate is Toffoli gate with polarity control. This gate has been used in several research papers for synthesizing logic algorithm. Previously proposed mixed-polarity Toffoli gate was not able to gain any attention because of low quantum anesthetization by use of NOT gate developed from Toffoli logic gate. Embodying Toffoli logic gate in the synthesis of reversible logic algorithm would increase the synthesis time and

S. C. Chua
Intel Corporation, Kuala Lumpur, Malaysia
e-mail: csc_chua@hotmail.com

A. K. Singh (✉)
Department of Computer Applications, NIT Kurukshetra, Kurukshetra, India
e-mail: ashutosh@nitkkr.ac.in

© Springer Nature Singapore Pte Ltd. 2020
A. K. Singh et al. (eds.), *Design and Testing of Reversible Logic*, Lecture Notes in Electrical Engineering 577, https://doi.org/10.1007/978-981-13-8821-7_6

also lack in terms of quantum cost and gate count improvement. The research work attraction enormously increases on this gate after synthesizing few of its quantum implementations. The use of this gate with quantum implementation gives better results. Now, it is able to compute complex reversible circuits with less quantum cost. Even though, many methods were proposed, but search-based reversible logic synthesis has never been discussed by using this gate in the available literature.

1.1   Literature Review

In this section, overview of the widely proposed reversible logic synthesis algorithms is discussed. The synthesis algorithms are being categorized into several subsections according to their types and functions.

1.1.1   Transformation-Based Synthesis Algorithm

The primitive developed algorithm of reversible logic synthesis is transformation-based synthesis algorithm. This synthesis algorithm was initiated by Iwama et al. [1] in which the author described various rules to convert Boolean function to a set of Toffoli-based gates. The introduced method shows that by using suggested rules, canonical form can be derived from any circuit. The proposed work was for conceptual concern only, unfortunately did not provide any practical illustration. This algorithm expanded by using number of predefined patterns (called templates) by appending transformation rules proposed by Miller et al. [2]. This algorithm operates by suggested transformation rules, which convert Boolean function to a set of Toffoli or CCNOT gates in the form of truth table. After getting the results, many elementary rules for template matching are applied for further simplification. The results obtained by using this synthesis algorithm were significant during that time. An extended version of the existing algorithm given by [1, 2] Dueck et al. [3] and Maslov et al. [4] which append their database by SWAP gate and Fredkin gate. The algorithm [2] refined later by Maslov et al. by launching bidirectional algorithm that repeatedly searching in the truth tables for getting the difference between input and output. This approach was different from conventional method that considered only the outputs. Many different techniques like MMD (Miller–Maslov–Dueck) [5], template matching, Reed–Muller spectra and resynthesis to synthesize reversible function in quantum-based cost and number of gates have been given by Malsov et al. [6]. This algorithm is not using the conventional representation of truth table that can be seen in the earlier algorithm and apply the Reed–Muller spectra form for reversible circuit operations derived from the library of NCT gate. By using Reed–Muller spectra, it provides greater gate substitution comparing to traditional approach and thus enhances the synthesis results. The flow of algorithm can be divided into three steps. First, MMD is applied that chooses a reversible circuit until it determines the entire identity term. Second, template matching is applied that used templates to remove the complexities of the synthesized network by selecting a small network. Third, resynthesis is applied for reducing the quantum cost by choosing a series of gates from the network.

1.1.2 Rule-Based Synthesis Algorithm

Arabzadeh et al. [7] suggested to use rule-based optimization technique to convert NCT (NOT, CNOT, Toffoli gate) to mixed-priority Toffoli-based circuit after applying transformation-based synthesis algorithm [2, 5]. The algorithm runs by supplementing rules to minimize the number of NOT gates in the circuit. The algorithm can be divided into two steps: In the first step, on giving reversible circuit, several NOT gates are applied and Toffoli-based circuit transforms into mixed-polarity Toffoli-based gate, and at this step, redundancy of NOT gates is removed. In the second step, a K-Map is used to find optimal gates by simplifying intermediate circuits.

Datta et al. [8] proposed a post-synthesis rule that uses template matching approach to convert NCT-based gate circuits to mixed-priority-based gates. The author proposed template matching rules for matching the cascade of various Toffoli-based circuits jointly to mixed-polarity-based circuits and replacing them. Their present template rules can match up to 4-variable mixed-polarity Toffoli-based gate.

1.1.3 Cycle-Based Synthesis Algorithm

Cycle-based synthesis algorithm divides the whole permutation into various cycles and each cycle synthesizes separately. The earliest cycle-based synthesis algorithm was proposed by Shende et al. [9] in which decomposition is applied on given permutation to transform into a set of disjoint cycles and synthesizing them separately. This technique works well when given permutations are not in regular patterns and reversible function, which result in several input combinations remain unchanged. Yang et al. [10] proposed a similar technique as of [9] in which besides selecting CNOT gate, it only choses NOT and Toffoli gates where qubid function for synthesizing has more than three variables. The algorithm introduces a number of permutation formula, which can easily be applied on functional area and convert them into synthesized gates. Prasad et al. [11] modified the existing algorithm [9] by replacing the use of Toffoli gate with NOT and CNOT gates in several conditions. By accepting these changes, results of gate count and quantum cost improves efficiently.

Sasanian et al. [12] improved the algorithm [9] by eliminating the decomposition of huge cycles into a set of 2-cycles which introduces a concept about utilizing reversible gates more. In the suggested method, huge cycles are decomposed into pair of 3-cycles, set of 3-cycles, and pair of 2-cycles and resulted cycles are synthesized directly. This implementation of cycles in the synthesis algorithm drastically reduced undesired reversible gates if compared to existing algorithm [9].

Saeedi et al. [13] proposed a k-cycle-based synthesis algorithm which uses a set of seven building blocks to synthesize a selected permutation to minimize the quantum cost of circuit and average runtime. Given a cycle of large inputs, the algorithm before synthesizing decomposes the rules to obtain a set of seven building blocks from the specified inputs. The seven building blocks contain a pair of 2-cycles, a single 3-cycles, a pair of 3-cycles, a single 5-cycles, a pair of 5-cycles, a single 2-cycles (4-cycles) followed by single 4-cycles(2-cycles). The algorithm is capable of synthesizing large variable reversible function (up to variable size of 20 with a time limit of 12 h per function) and it gives better results in terms of quantum cost for reversible benchmark function of variable size 8 and above.

Saeedi et al. [14] introduced an algorithm which utilizes the set of building blocks and library for synthesizing given instructions. Each instruction is considered as a permutation with various cycles where few reversible gates are applied for the synthesis of each cycle. The experimental results of proposed methods minimized the quantum cost and average runtime.

1.1.4 Binary Decision Diagrams-Based Synthesis

Binary Decision Diagrams (BDDs) based synthesis algorithm was initiated by Kerntopf et al. [15]. Decision diagram is constructed from any Boolean function by considering all possible gates. The gates that minimize the complexity of Decision diagrams are selected. A similar process is repeated for further analysis. The process terminates when all the paths are analyzed and choose the minimal path of node and obtain reversible circuits. Wille et al. [16] proposed an algorithm which begins by constructing BDD. Every BDD node is replaced by a number of reversible circuits. In BDDs, shared nodes may be present, it results in fan-outs which is not granted in reversible logic. To prevail over additional ancilla/constant, input and CNOT gates are applied. Even though algorithm helps to reduce the quantum cost but a large number of ancilla/constant inputs usage is not feasible. The algorithm is upgraded later by Wille et al. [17], by proposing a post-processing optimization method that reduces the garbage line by adding some garbage output lines, which is suitable with ancilla input lines. The problem still exists even after so many improvements.

Krishana et al. [18] proposed a method, which uses isomorphic subgraph matching that helps in mapping BDD subgraph to reversible circuit template structure. By using this method, it drastically reduces ancilla/constant inputs made by existing BDD-based synthesis algorithm [16]. The problem of ancilla BDD-based synthesis algorithm was solved by recently proposed synthesis algorithm given by Soeken et al. [19]. The proposed concept symbolic function is implemented very efficiently with BDDs, which is based on young subgroups that use symbolic function.

1.1.5 Search-Based Synthesis Algorithm

Search-Based Synthesis Algorithm proposed by Gupta et al. [20], which used positive-polarity Reed–Muller (PPRM) expansion to synthesize reversible logic in the form of gate count. This algorithm implemented by traverse a search tree for input reversible function and by using gate matching factor determines all feasible solutions. In a given period, the algorithm traversed all feasible solutions and store the optimal solution.

Later, Saeedi et al. [21] improved the algorithm by proposing various new substitutions rules to the fundamental ones. Their invented results show that method conducts better synthesis of circuit in terms of total cost and also the probability of leading a synthesized circuit in a better way as compared to existing algorithm [20].

This algorithm extended by Donald et al. [22] for dealing with SWAP gates, Peres gate, Fredkin gate, and reverse Peres gate in a similar domain of search-based framework. In this algorithm potential of circuit synthesis in term of quantum cost is also added while keeping the synthesis approach same as of [20].

1.1.6   Graph-Based Synthesis Algorithm

Graph-Based Synthesis Algorithm is introduced by Yexin et al. [23] that deals with the synthesis of reversible logic in set of gate count using NCT-based library gates. In each synthesis step, the algorithm focus on fasting synthesis time by minimizing functional complexity. Even though synthesis time is reduced, the algorithm is able to find all possible solutions for every input reversible function. The disadvantage of algorithm is that in each step, it is unable to determine optimal gate for substitution and results as unnecessary use of gates to synthesize a function.

1.1.7   Optimal-Based Synthesis Algorithm

Golubitsky et al. [24], the author proposed an Optimal-Based Synthesis Algorithm that is able to find the optimal solution by matching all input reversible functions into a set of presents in database. By use of database, the gate count of optimal circuit can be found in lesser time through searching in their canonical representative form. The author [25] extended this approach for improvement in existing database [24] by removing all reverse functions and inserted various new functions to existing database. This upgrades the algorithm performance. Now, the algorithm is able to synthesize higher variables reversible functions. When handling large reversible function, the algorithm breaks the function into a number of sub-functions using database individually, matching them, and finally, combine them.

Szprowski et al. [26] extended the synthesized algorithm by adding method for further synthesis in form of quantum cost [24]. The introduced algorithm can generate optimal circuits based on reversible functions up to four variables. Even though gate count is optimal but not assuring that these are minimal circuits in quantum-based cost. The reason behind it, there may exist several longer cascades with lesser value of quantum cost. In Golubitsky et al. [25], the author upgrades an algorithm [26] by utilizing classical decomposition method without requiring additional lines. By using this approach, it provides significant savings of quantum cost. This algorithm

[27] further improved by Szyprowski et al. [28], by connecting mixed-polarity-based Toffoli gate to the synthesis library.

Li et al. [29] introduced an algorithm which is unable to find optimal solution up to 4-bit reversible function by using NCTP library (NOT, CNOT, Toffoli and Toffoli4, Peres, and inverse Peres gates). When synthesizing a given function, primarily, the introduced algorithm uses a hash table to search a combination with minimal gate count by using NOT, CNOT, Toffoli, and Toffoli4 gates. Then the split hash tables are merged to produce a longer has table by applying several computation rules. Lastly, this algorithm synthesizing the resulted circuit in quantum cost terms by transforming the selected path to Peres and inverse Peres gates. The introduced algorithm designed a memory in an efficient way to use lesser amount of coding for computing permutation rules and topological compression are applied. It also uses flexible data structures for memory savings. As compared to existing synthesis algorithm [26], this algorithm produced better results in term of synthesis time and quantum cost.

1.2   Contribution of This Chapter

Search-based reversible logic synthesis algorithm is shown to be capable of putting to good use of reversible logic better as compared to available synthesizing algorithms. The available search-based synthesis algorithm is proposed by Gupta et al. and Donald et al. They already proved that for realizing any reversible function, the algorithm is capable to find all feasible realizations of reversible logic circuits. Search-based synthesis algorithm does not generate any extra lines similar to BDD-based realization algorithm. Thus, it considers the synthesized function in its simplest form with least amount of ancilla inputs and garbage outputs. The best part of this algorithm is that it is not demanding for a blend of dissimilar approaches as transformed-based synthesis algorithm. For additional usage, the produced realization solution is saved in database of optimal-based synthesis algorithm.

A proposal of search-based synthesis algorithm using mixed-polarity Toffoli gate is given in this chapter. The use of NCT library with mixed-polarity Toffoli gate for search-based reversible logic synthesis algorithm has been also introduced. Mixed-polarity Toffoli gate for search-based reversible logic synthesis algorithm is discussed for the first time in literature. The reversible function is synthesized by introducing algorithm in positive-polarity Reed–Muller expansion (PPRM) and choose appropriate gates for replacement by searching on matched sets in the expansion. This algorithm focuses on quantum cost and number of gate count required for the synthesis of reversible logic. The algorithm keeps minimal garbage output by not generating ancilla input.

2   Conversion of Logic Function into PPRM Expansion

Before describing the approach of the introduced synthesis algorithm, a prime matter
is discussed on irreversible logic function transforming to PPRM expansion which is
set as input to the introduced algorithm. The first step is to convert the nonreversible
function to reversible. This is achieved by adding extra lines to perform a 1:1 mapping
among the input and the output. When reversible function diagnosed, the next step
is transformed into ESOP expansion.

Here, those functions are performed by EXORCISM-4 program [30]. This program is used by Boolean function for heuristics and efficient look-ahead strategies to
search the ESOP expansion quickly. After acquiring the ESOP expansion, transform
it into PPRM expansion. This is obtained by applying rules $\overline{x_n} = 1 \oplus x_n$ to inverted
variables and reduction is obtained by canceling out identical variables even numbers of $x_n \oplus x_n = 0$. In last, PPRM expansion is achieved. In this work, the logic
function is achieved by PPRM expansion by using the following steps:

1. The irreversible functions are translated in a manual manner to reversible function by applying additional required lines. Then, EXORCISM-4 is applied for
   changing them in ESOP expansion. Later that expansion is translated to PPRM
   expansion. This is done by using C language and the solutions are saved in PPRM
   expansion (.pprm).
2. Many reversible benchmark functions are gathered by [28, 31, 32]. They are
   available in ".blif" or ".pla" expansion. Then, these are changed to the PPRM
   expansion by applying a program and EXORCISM-4.

3   Synthesis Algorithm

3.1   Gate Substitutions

The given reversible function is converted into PPRM expansion, and the initiated
algorithm first finds the expansion and searches all possible matching reversible gates.
A new path is created for every selected reversible gate and that gate is replaced in
the native task to obtained reduced form of PPRM expansion. For determining the
reversible gate which is to be selected for high probability better solution, definite
requirements are set. When requirements are fulfilled, reversible gates get selected.
The substitution and requirements of every reversible gate are in the following ways:

For the selection of NCT gate library, only if the PPRM expansion set holds a
single variable term xi and $factor$, where $factor$ termed as any single variable
to several multiple variable terms that does not to be formed of xi. In case of NOT
gate, the $factor$ termed as constant one. An exchanged of $xi \oplus factor$ is
created on expansion holding variable xi. The chosen gates in the task are bounded
by the amount of variables in a manner that the largest gate must not be greater than

the term of task variable. For example, when a 3- variable-based functions, Toffoli gate will be chosen as the greatest gate and for synthesizing four variable functions, the selected greatest gate will be Toffoli-4.

The m-NCT gate library will selected by $n - 2$ negative line (for n line variable) if and only if the term of PPRM expansion contain any single term variable of xi and $factor$, where $factor$ is termed as any other two variable terms to several multiple variable terms that does not to be formed of xi. Then substitution of gate will be executed on expansion holding variable term of xi, from the $factor$ term selection of negative control variable is done.

An m-NCT gate library will be selected with $n - 1$ negative line (for n line variable) if and only if the term of PPRM expansion contain any single term variable of xi and $factor$, where $factor$ is termed as any other two variable term to several multiple variable terms that does not to be formed of xi. Then substitution of gate will be executed on expansion holding variable term of xi, from the $factor$ term selection of negative control variable is done.

3.2 Hamming Distance

While reversible function from one path reduced to another path, the method will search for all feasible reversible gates which are able to simplify the native function. The results of single PPRM expansion may have several replacements of different gates. Every substitution is stored in a separate path. To find the path that will give a fast and better solution, Hamming Distance (HD) method [2] is used. This method determines the quantity of variables contrast among the input and output terms. On the basis of experiment results, the paths that hold minimum not equal to 0 HD has the largest possibility to lead speedy for better results. So, the track holding minimum not equal to 0 HD chosen as the next converting track.

The HD of a corresponding path becomes zero, when a solution is verified. Information about the path will be stored in terms of gate count, gate connection, and quantum cost. The method pursues to find the next minimum not equal to 0 HD path having the potential for best results. When all paths are traversed, then this method will stop. This improves the memory and synthesis time of the algorithm, and the program sets to abort its path before the result is determined. That can be achieved by differentiating the present quantum cost and gate count gathered with the present optimal result on the path. The particular path is terminated, if it fails to generate better results.

3.3 Algorithm

The function inputs are reversible function expressed in its PPRM expansion $f(x1, x2, \ldots, xn)$. Depending upon receiving the input in the form of PPRM expansion, this algorithm initialize execution and initially assign zero to quantum cost and gate count. During this stage, the HD expansion is computed. Then, algorithm finds the PPRM expansion and checks all feasible matching in reversible gates for more reduction. After determining the reversible gate, replacement will be achieved and assigned path with a new number. Current path updated corresponds to gate count, quantum cost, and HD. Further, the algorithm chooses the next synthesize path by picking up a path that contains minimum HD which is not equal to 0. This algorithm iterates till a feasible solution is reached, and the calculated HD of path is close to 0. The whole data of the path about the usage of reversible gate, connections of the whole circuit, assembled gate countm and quantum cost are stored as an optimal possible solution. For further differentiation, a duplicate assembled gate count and quantum cost are stored in register $best\_gate\_count$ and $best\_quantum\_cost$. The algorithm then pursues for searching the next minimum HD path with not equal to 0. At any time, a gate replacement is formed and a new path is developed, then synthesis algorithm determines the result according to the ongoing path. If the path not generating better results as compare to current optimal results then terminate that path and gathered data is also disposed of. The comparison is done in a given way:

When realizing in set of gate counts, termination of path if $gate\_count(path) > best\_gate\_count - 1$ or $gate\_count(path) == best\_gate\_count - 1$ with $quantum\_count \geq best\_quantum\_cost - 1$. As for synthesizing in terms of quantum cost, termination of path if $Quantum\_cost(path) > best\_quantum\_cost - 1$ or $gate_count(path) == best\_gate\_count - 1$ with $gate\_count(path) \geq best\_gate\_count - 1$. For those cases where the found path provides a better solution at HD $= 0$, the replacement of ongoing optimal result take place. This algorithm must meet stopping criteria after exploring all paths or the timer given by the user has been reached (Fig. 1).

3.4 Example

For perception of the introduced algorithm, it is illustrated by taking an example of reversible function by synthesizing with detail explanation of every step, then describes the achievement of optimal solution. In the given illustration, when algorithm finds a path, some reversible gates are chosen and checks matching of all feasible gates. The benchmark function selected for demonstration is named as "3_7" and having features [0,1,2,3,4,5,6,7] [31]. Then stated in PPRM expansion, it provides $a' = 1 \oplus a \oplus c \oplus ab \oplus ac$, $b' = 1 \oplus a \oplus b \oplus c$, $c' = 1 \oplus a \oplus b \oplus ab \oplus bc$. In tabular format, the expression of program are stored, as given in Table 1. The HD calculated is 13 for the given path.

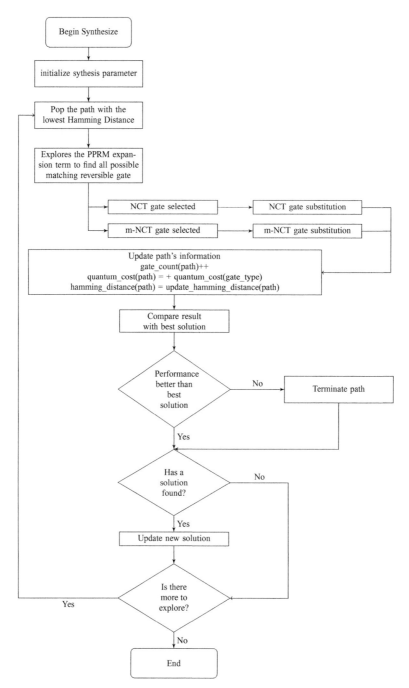

Fig. 1 Flow chart of function Synthesize()

Table 1 Comparison table of all discussed synthesis methods

Synthesis method	Features	Limitation	Library	Metric
Iwama et al. [1]	• Transformation-based synthesis	• Large amount of ancilla inputs and garbage output • Circuit dependency	NCT	GC
Miller et al. [2]	• Transformation-based synthesis • No ancilla input	• Limited Scalability • Circuit dependency	NCT	GC
Dueck et al. [3]	• Transformation-based synthesis	• Large amount of ancilla inputs and garbage output • Circuit dependency	NCTSF	GC
Maslov et al. [4]	• Transformation-based synthesis • No ancilla input	• Limited Scalability • Circuit dependency	NCTSF	GC
Maslov et al. [5]	• Transformation-based synthesis • No ancilla input	• Limited Scalability	NCT	GC
Maslov et al. [6]	• Transformation-based synthesis • Able to cope with large function	• Limited Scalability	NCT	GC
Arabzadeh et al. [7]	• Rule-based synthesis • No ancilla input	• Circuit dependency	m-NCT	QC
Datta et al. [8]	• Rule-based synthesis • No ancilla input	• Limited scalability • Circuit dependency	m-NCT	GC, QC
Shende et al. [9]	• Cycle-based synthesis • No ancilla input	• Circuit dependency • Limited scalability	NCT	GC
Yang et al. [10]	• Cycle-based synthesis • No ancilla input	• Circuit dependency • Limited scalability	NCT	GC
Prasad et al. [11]	• Cycle-based synthesis • No ancilla input	• Circuit dependency • Limited scalability	NCT	GC
Sasanian et al. [12]	• Cycle-based synthesis • No ancilla input	• Limited scalability	NCT	GC
Saeedi et al. [13]	• Cycle-based synthesis • No ancilla input • Able to cope with large function	• Circuit dependency	NCT	QC
Saeedi et al. [14]	• Cycle-based synthesis • No ancilla input • Able to cope with large function	• Circuit dependency	NCT	QC
Kerntopf [15]	• BDD-based synthesis	• Limited scalability	NCTSF	QC
Wille and Drechsler [16]	• BDD-based synthesis • Able to cope with large function	• Ancilla input • Garbage output	NCT	QC
Wille et al. [17]	• BDD-based synthesis • Able to cope with large function	• Ancilla input • Garbage output • Circuit dependency	NCT	Ancilla input

(continued)

Table 1 (continued)

Synthesis Method	Features	Limitation	Library	Metric
Krishna and Chattopadhyay [18]	• BDD-based synthesis • Able to cope with large function up to 9 variables • Much efficient in handling	• Ancilla input • Garbage output • Poor scalability	NCT	Ancilla input
Krishna and Chattopadhyay [18]	• BDD-based synthesis • Able to cope with large function up to 9 variables • Much efficient in handling ancilla input	• Ancilla input • Garbage output • Poor scalability	NCT	Ancilla input
Socken, et al. [19]	• DDD-based synthesis • Able to cope with large function • Ancilla free	• Circuit dependency • Requires large memory space	NOT	QC, Ancilla input
Gupta, et al. [20]	• Search-based synthesis	• Limited scalability	NCT	GC
Saeedi, et al. [21]	• Search-based synthesis	• Limited scalability	NCT	QC
Donald and Jha [22]	• Search-based synthesis	• Limited scalability • Slow synthesis time	NCTSFP	GC, QC
Yexin et al. [23]	• Graph-based synthesis • Fast synthesis time	• Poor scalability	NCT	GC
Golubitsky et al. [24]	• Optimal-based synthesis • Fast synthesis time	• Function dependency • Limited scalability • Requires large amount of storage	NCT	GC
Golubitsky et al. [25]	• Optimal-based synthesis • Fast synthesis time • Able too cope with large function	• Function dependency • Limited scalability • Requires large amount of storage	NCT	GC
Szyprowski et al. [26]	• Optimal-based synthesis • Fast synthesis time	• Function dependency • Limited scalability	NCT	GC
Szyprowski et al. [27]	• Optimal-based synthesis • Fast synthesis time	• Function dependency • Limited scalability	NCT	GC, QC
Szyprowski et al. [28]	• Optimal-based synthesis • Fast synthesis time	• Function dependency • Limited scalability	m-NCT	GC, QC
Li et al. [29]	• Optimal-based synthesis • Fast synthesis time	• Function dependency • Only limited to 4-variable-based reversible function	NCT	QC

Table 2 Benchmark function given in PPRM expansion

Coefficient	Function
	aaa
1	111
a	111
b	011
ab	101
c	110
ac	100
bc	001
abc	000
HD	13
Gate applied	–
GC (QC)	0 (0)

The PPRM expansion terms are analyzed by algorithm and find all possible reversible gates which are matched. For each selected gate, substitution is performed and generate functions, which are new and are saved as sub-paths. Assigned distinctive numbers to path and information is updated about gate connection, quantum cost, and gate count. Table 2 shows four results of sub-paths with respect to substitution of gates. At row named as "gate applied", the character "1" denotes control line of gate; "x" denotes gate target line; "–" denotes no connection, and "o" denotes negative line control (for library gate of m-NCT).

In Table 2, from the transformed paths, for every selected gate, HD is calculated by this algorithm. As not even a single path has zero HD, it shows valid solution that has been found at this level. Then, this algorithm selects the nonzero minimum HD path as its succeeding synthesized path. The path having minimum nonzero HD is shown by a shaded color. The gates get selected and substitution is performed on this path and their results are given in Table 3. The updation can also be seen in terms of gate count, HD, and quantum cost for new path.

The minimum nonzero HD is 7, as shown in Table 3. On the path, more reduction is performed and results displayed in Table 4.

The minimum nonzero HD is 4, as shown in Table 4. More reduction is performed on the path and is shown in Table 5. It is observed that the output line b reach to its identical term, so more depletion on path b and use of this control line has no significance.

The minimum nonzero HD is 2, as shown in Table 5. More reduction is performed on the path and results are given in Table 6.

In Table 6, the shaded color shows HD path reached zero, and it indicates a valid solution which is caught by the path data. The remaining paths which have the path similar to parent not close to HD zero, exploring on these paths will not give a better solution. Therefore, their whole information is discarded. The reason behind that,

Table 3 After first gate substitution

Coefficient	Function			
	abc	abc	abc	abc
1	111	000	100	001
a	111	111	010	011
b	011	110	011	010
ab	101	101	101	101
c	110	010	110	110
ac	100	100	100	100
bc	001	001	001	001
abc	000	000	000	000
HD	11	9	11	11
Gate applied	TOF x–1	TOF x–	TOF–x–	TOF–x
GC (QC)	1 (1)	1 (1)	1 (1)	1 (1)

Table 4 After second gate substitution

	Function			
Coefficient	abc	abc	abc	abc
1	000	000	000	000
a	111	111	100	110
b	110	110	110	111
ab	101	101	101	101
c	010	001	010	100
ac	100	100	101	001
bc	111	100	001	001
abc	000	000	000	000
HD	11	7	8	9
Gate applied	TOF x11	TOF x–1	TOF–x1	TOF–x1
GC (QC)	2 (6)	2 (3)	2 (3)	2 (3)

these paths will require more reduction to reach a solution, means it needs additional gate count and never gives a better solution. Therefore, these path discarded by the algorithm and processes to traverse the upcoming minimum nonzero HD path to search for the best solution.

Table 5 After third gate substitution

Coefficient	Function			
	abc	abc	abc	abc
1	000	000	000	000
a	111	111	100	010
b	110	100	110	110
ab	101	101	101	000
c	001	001	001	001
ac	100	100	000	100
bc	010	110	100	100
abc	000	000	000	000
HD	7	9	4	5
Gate applied	TOF x11	TOF x1o	TOF 1x-	TOF 1ox
GC (QC)	3 (8)	3 (8)	3 (4)	3 (8)

Table 6 After fourth gate substitution

Coefficient	Function		
	abc	abc	abc
1	000	000	000
a	100	101	100
b	110	110	110
ab	101	100	000
c	001	000	001
ac	000	000	000
bc	101	100	100
abc	000	000	000
HD	5	6	2
Gate applied	TOF x11	TOF x1o	TOF 11x
GC (QC)	4 (9)	4 (9)	4 (9)

4 Synthesis Result

The results of the introduced algorithm are presented in this section. All experiments were performed on a Dell Precision T1600 featured an Intel Xeon CPU E31280 processor at 3.5 GHz, 512 kB cache, 16 RAM, and running on Windows 7 Professional edition. The proposed algorithm has been implemented using C language running on Ubuntu 10 through Oracle VM VirtualBox.

4.1 Three-Variable-Based Reversible Functions

To select an efficient synthesis-based reversible algorithm, a method is used to observe the results for all 3-variable-based reversible functions. The introduced algorithm for synthesizing results of all three reversible-based functions is given in this section, which contains 40,320 total functions. Quantum cost and gate count for both m-NCT and NCT gate libraries are calculated. These results are further compared with other proposed algorithms given by different authors. Among the best algorithm classified according to algorithmic paradigm (like cycle-based, search-based, transformation-based, and BDD-based) are chosen. The synthesized results in terms of gate count of proposed algorithm by all three-based reversible functions with comparison to other algorithms is shown in Table 7. The synthesized circuits are classified according to the number of gate counts shown in column "Number of Gates". The details given in other columns show the circuit numbers synthesized by denoted quotations. The minimum number of gate counts to synthesis any reversible function is the ideal case. The results calculated by the synthesis algorithm are given in "Proposed Work" column. In the given table, the introduced algorithm results are in contrast with available synthesis algorithm [9]. It can be analyzed that the proposed results using NCT library gate are near to best solution and improved as compared to many previous synthesis algorithm [5, 22, 23]. It gives a better solution than the currently available best search-based synthesis algorithm described by Donald et al. [22]. The significant decrease in number of gate count after including m-NCT library gate. Overall, this algorithm generates minimum average gate count and any circuit does not require more than 8 gates.

In spite of synthesizing in form of gate count, the introduced algorithm is also able to synthesis reversible function on the basis of quantum cost. The results of introduced

Table 7 After fourth gate substitution

Coefficient	Function		
	abc	abc	abc
1	000	000	000
a	100	101	100
b	110	110	110
ab	101	100	000
c	001	000	001
ac	000	000	000
bc	101	100	100
abc	000	000	000
HD	5	6	2
Gate applied	TOF x11	TOF x1o	TOF 11x
GC (QC)	4 (9)	4 (9)	4 (9)

Table 8 Gate count for all 3-variable-based reversible functions

Number of gates	Proposed work NCT	Proposed work m NCT	Donald et al. [22] NCT	Maslov et al. [5] NCT	Maslov et al. [6] NCT	Yexin et al. [23] NCT	Optimal solution Shende et al. [9] NCT
0	1	1	1	1	1	1	1
1	12	21	12	12	12	12	12
2	102	225	102	102	102	90	102
3	625	1527	625	567	625	476	625
4	2702	6058	2642	2125	2780	1833	2780
5	7932	14139	7479	5448	8819	4996	8921
6	14384	14995	13596	9086	16953	9126	17049
7	12201	3273	12476	9965	10367	10630	10253
8	2339	81	3351	7274	659	7820	577
9	22	0	36	3837	2	3788	0
10	0	0	0	1444	0	1265	0
11	0	0	0	391	0	258	0
12	0	0	0	62	0	25	0
13	0	0	0	6	0	0	0

algorithm in terms of quantum cost for all 3-variable reversible functions are shown in Table 8. The synthesized circuit are classified by quantum cost size are given in Table 8. The other columns data denotes the number of reversible functions that come in the classification on size of available quantum cost. As any other work does not have quantum cost table so unable to perform comparison test. Although, these generated synthesis results guaranteed to be near optimal solution. This only happened because the introduced algorithm examines and confirms the produced quantum cost of each subsection and chooses only the path which leads to the minimum quantum cost. In addition, it helps to improve quantum cost results by the use of m-NCT library.

When synthesized in terms of gate count, the best feature of the algorithm provides the lowest quantum cost. This assured that the optimal quantum cost exists for corresponding every gate count. The same is applicable for minimum gate count when synthesized in terms of quantum cost. The comparison of introduced algorithm with the characteristic turn on and off when synthesis in terms of gate count using NCT library gate is shown in Table 9. It is observed that the appended characteristics provide an improved result.

Table 9 Comparison of gate count and quantum cost

Quantum cost size	Proposed work NCT	Proposed work m-NCT
0	1	1
1	9	9
2	51	51
3	187	187
4	387	387
5	426	432
6	305	353
7	350	560
8	1305	1812
9	2952	3458
10	3418	2938
11	1416	1001
12	946	1964
13	3543	5728
14	7278	7851
15	6095	2798
16	1017	856
17	950	2601
18	3319	5048
19	4884	2221
20	1461	64
21	20	0

4.2 Benchmark Function

All functions are benchmark which is obtained from [28, 31, 32]. The introduced algorithm usage to synthesizing benchmark functions on the basis of quantum cost and gate count results is shown in Table 10. The tables are arranged according to size of benchmark reversible functions. All shown results are optimal determined result from the introduced algorithm by holding minimal quantum cost or gate count.

The arrangement of tables is done in the given way. The row of table shows the benchmark name and function's size. The succeeding rows are created with variety of columns, which shows the experimental results corresponding to benchmark function. The initial column displays the library gate operate to manage the simulation. The upcoming two columns displays the result of quantum cost and gate count. The succeeding column holds the benchmark circuit connection decided by the algorithm.

The notation follows the following circuit connection format $tx \ b_1, \ldots, b_n, b_{n+1}$ where t indicates Toffoli gates, x indicates the size of Toffoli gate and $b_1, \ldots, b_n, b_{n+1}$

Table 10 Quantum cost result for all 3-variable-based reversible functions

Gate count/quantum cost	Synthesis in term of GC with QC minimization		Synthesis in term of GC without QC minimization	
	GC	QC	GC	QC
0	1	1	1	1
1	9	9	9	9
2	51	51	51	51
3	187	187	187	187
4	387	387	387	387
5	426	426	426	426
6	305	305	305	305
7	350	350	350	347
8	1305	1305	1305	1267
9	2952	2946	2952	2753
10	3418	3488	3418	2981
11	1416	1377	1416	1088
12	946	937	946	915
13	3543	3543	3543	3483
14	7278	7246	7278	6847
15	6095	5945	6095	5176
16	1017	949	1017	754
17	950	980	950	1173
18	3319	3358	3319	3765
19	4884	4893	4884	5177
20	1416	1463	1416	1478
21	20	62	20	122
22	0	138	0	537
23	0	74	0	846
24	0	0	0	233
25	0	0	0	0
26	0	0	0	0
27	0	0	0	5
28	0	0	0	7

are control line coefficients and b_{n+1} is the target bit. If a negative control line coefficient is detected, then a stroke will be given. For example, Toffoli4 with a, c, d control lines, d is a negative control line and b as target line, the written notation is $t4\ a, c, d', b$.

5 Summary

The synthesis algorithm that uses the NCT library gate with mixed-polarity control is presented. The introduced synthesis algorithm gives reversible function by using the PPRM expansions and then HD method is applied to choose the transformation path. For the chosen path, several reversible gates are determined by searching feasible reversible gate matching. The introduced algorithm keeps the minimal garbage output as it does not produce extra ancilla input. This algorithm has been used to synthesize all 3- variable-based reversible function and the significant results are obtained. It provides the usage of introduced algorithm to synthesize the reversible benchmark functions. From the experimental results, it has been provided that by adding m-NCT library, the results are greatly improved.

References

1. Iwama K, Kambayashi Y, Yamashita S (2002) Transformation rules for designing CNOT-based quantum circuits. In: Proceedings of the 39th annual design automation conference, New York, pp 419–424
2. Miller DM, Maslov D, Dueck GW (2003) A transformation based algorithm for reversible logic synthesis. In: Proceedings of the 40th annual design automation conference, New York, pp 318–323
3. Dueck GW, Maslov D, Miller DM (2003) Transformation-based synthesis of networks of Toffoli, Fredkin gates. In: Canadian conference on 2003 electrical and computer engineering, IEEE CCECE, pp 211–214
4. Maslov D, Dueck GW, Miller DM (2005) Synthesis of Fredkin-Toffoli reversible networks. IEEE Trans Very Large Scale Integr (VLSI) Syst 13:765–769
5. Maslov D, Dueck GW, Miller DM (2005) Toffoli network synthesis with templates. IEEE Trans Comput-Aided Des Integr Circuits Syst 24:807–817
6. Maslov D, Dueck GW, Miller DM (2007) Techniques for the synthesis of reversible Toffoli networks. ACM Trans Des Autom Electron Syst 12
7. Arabzadeh M, Saeedi M, Saheb Zamani M (2010) Rule-based optimization of reversible circuits. In: Proceedings of the 2010 Asia and South Pacific design automation conference, New Jersey, pp 849–854
8. Datta K, Rathi G, Wille R, Sengupta I, Rahaman H, Drechsler R (2013) Exploiting negative control lines in the optimization of reversible circuits. In: Reversible computation. Springer, Berlin, pp 209–220
9. Shende VV, Prasad AK, Markov IL, Hayes JP (2003) Synthesis of reversible logic circuits. IEEE Trans Comput-Aided Des Integr Circuits Syst 22:710–722
10. Yang G, Song X, Hung WN, Xie F, Perkowski M (2006) Group theory based synthesis of binary reversible circuits, pp 365–374
11. Prasad AK, Shende VV, Markov IL, Hayes JP, Patel KN (2006) Data structures and algorithms for simplifying reversible circuits. ACM J Emerg Technol Comput Syst 2:277–293
12. Sasanian Z, Saeedi M, Sedighi M, Zamani MS (2009) A cycle-based synthesis algorithm for reversible logic. In: Design automation conference, pp 745–750
13. Saeedi M, Saheb Zamani M, Sedighi M, Sasanian Z (2010) Reversible circuit synthesis using a cycle-based approach. ACM J Emerg Technol Comput Syst 6
14. Saeedi M, Sedighi M, Saheb Zamani M (2010) A library-based synthesis methodology for reversible logic. Microelectron J 41:185–194

15. Kerntopf P (2004) A new heuristic algorithm for reversible logic synthesis. In: Proceedings of 41st design automation conference, Poland, pp 843–837
16. Wille R, Drechsler R (2009) BDD-based synthesis of reversible logic for large functions. In: Proceedings of the 46th annual design automation conference, New York, pp 270–275
17. Wille R, Soeken M, Drechsler R (2010) Reducing the number of lines in reversible circuits. In: Proceedings of the 47th design automation conference, New York, pp 647–652
18. Krishna M, Chattopadhyay A (2014) Efficient reversible logic synthesis via isomorphic subgraph matching. In: 2014 IEEE 44th international symposium on multiple-valued logic (ISMVL), pp 103–108
19. Soeken M, Tague L, Dueck GW, Drechsler R (2016) Ancilla-free synthesis of large reversible functions using binary decision diagrams. J Symb Computat 73:1–26
20. Gupta P, Agrawal A, Jha NK (2006) An algorithm for synthesis of reversible logic circuits. IEEE Trans Comput-Aided Des Integr Circuits Syst 25:2317–2330
21. Saeedi M, Saheb Zamani M, Sedighi M (2007) On the behavior of substitution-based reversible circuit synthesis algorithm investigation and improvement, Washington, pp 428–436
22. Donald J, Jha NK (2008) Reversible logic synthesis with Fredkin and Peres gates. ACM J Emerg Technol Comput Syst 4
23. Yexin Z, Chao H (2009) A novel Toffoli network synthesis algorithm for reversible logic. In: Design automation conference, ASP-DAC 2009. Asia and South Pacific, pp 739–744
24. Golubitsky O, Falconer SM, Maslov D (2010) Synthesis of the optimal 4-bit reversible circuits. In: Proceedings of the 47th design automation conference, pp 653–656
25. Golubitsky O, Maslov D (2012) A study of optimal 4-bit reversible Toffoli circuits and their synthesis. IEEE Trans Comput 61:1341–1353
26. Szyprowski M, Kerntopf P (2011) Reducing quantum cost in reversible Toffoli circuits. arXiv:1105.5831
27. Szyprowski M, Kerntopf P (2011) An approach to quantum cost optimization in reversible circuits. In: 2011 11th IEEE conference on nanotechnology (IEEE-NANO), pp 1521–1526
28. Szyprowski M, Kerntopf P (2013) Optimal 4-bit reversible mixed-polarity Toffoli circuits. In: Glück R, Yokoyama T (eds) Reversible computation, vol 7581, pp 138–151. Springer, Berlin
29. Li Z, Chen H, Song X, Perkowski M (2014) A synthesis algorithm for 4-bit reversible logic circuits with minimum quantum cost. ACM J Emerg Technol Comput Syst (JETC) 11:29
30. Mishchenko A, Perkowski M (2001) Fast heuristic minimization of exclusive-sums-of-products. In: International workshop on applications of the reed-muller expansion in circuit design, pp 242–250
31. Maslov D, Dueck G, Scott N (2005) Reversible logic synthesis benchmarks page. http://www.cs.uvic.ca/~dmaslov
32. Wille R, Große D, Teuber L, Dueck GW, Drechsler R (2008) RevLib: an online resource for reversible functions and reversible circuits. In: 2008 38th international symposium on multiple valued logic, ISMVL 2008, pp 220–225

Reversible Circuit Synthesis Using Evolutionary Algorithms

T. N. Sasamal, H. M. Gaur, A. K. Singh and A. Mohan

Abstract With the unprecedented growth in VLSI technology in recent years, managing power dissipation has become a challenging task for many researchers. In this aspect, reversible logic emerges as one of the basis of future lossless computing system that promises zero energy dissipation, meanwhile classical physics cannot survive due to constant scaling of transistors and the exponential growth of transistor density in integrated circuits. It has applications in various domain such as low power VLSI, fault-tolerant designs, quantum computing, nanotechnology, DNA computing, optical computing, cryptography, and informatics. There are many existing works for the synthesis of reversible logic circuits; some are exact methods while others based on heuristic approaches. In this survey, we review a range of evolutionary computation approaches to the problem of optimal synthesis of reversible Logic—GA (Genetic Algorithm) based, PSO (Particle Swarm Optimization) based, ACO (Ant Colony Optimization)-based circuits where aim is to obtain a near-optimal solution by efficiently exploring the entire search space. This study provides an algorithmic review with comparative study on metaheuristic-based reversible logic synthesis methods proposed in existing literatures. Comparison of experimental results based on large number of benchmark circuits conform that evolutionary algorithms-based technique enables optimal or near-optimal solutions with lesser synthesis time.

T. N. Sasamal (✉)
Department of Electronics & Communication Engineering,
NIT Kurukshetra, Kurukshetra, India
e-mail: tnsasamal.ece@nitkkr.ac.in

H. M. Gaur
Department of Electronics & Communication Engineering, ABES Institute of Technology,
Ghaziabad (Delhi NCR), India
e-mail: leoharimohan84@gmail.com

A. K. Singh
Department of Computer Applications, NIT Kurukshetra, Kurukshetra, India
e-mail: ashutosh@nitkkr.ac.in

A. Mohan
Department of Electronics Engineering, IIT BHU, Varanasi, India
e-mail: profanandmohan@gmail.com

© Springer Nature Singapore Pte Ltd. 2020   115
A. K. Singh et al. (eds.), *Design and Testing of Reversible Logic*, Lecture Notes
in Electrical Engineering 577, https://doi.org/10.1007/978-981-13-8821-7_7

1 Introduction

In the current scenario with increasing complexity in VLSI circuits; managing power dissipation is an important issue in digital circuit design. With higher levels of integration and increasing scaling; Moore's law seems to be valid yet, but in traditional (irreversible) technologies heat produced by each IC doubles [44]. Lossless computing offers an alternative, where logical operations do not yield information loss called reversible operation [36, 69]. Reversible logic realizes n-input n-output functions where a bijective relation exists between input and output vector. In a reversible logic, every input pattern can be uniquely recovered from its output pattern, so no information is lost during computation. Reversible logic circuits take care of heat loss due to information erase. Thus reversibility will become an inherent property that will help to broaden low power design [66] and quantum computation [23, 46] horizon, and also have applications to fault-tolerant designs, nanotechnology, DNA computing, optical computing, cryptography, and informatics [36, 37, 46]. Work by Landauer [22] showed that, regardless of the underlying technology, Conventional logic circuits dissipate heat in an order of kTln2 joules for every bit of information that is lost, where k is the Boltzmann constant and T is the operating temperature. Since today's computing devices are usually built of elementary gates like AND, OR, NAND, etc., they are subject to this principle and, hence, dissipate this amount of power in each computational step.

Synthesis of reversible logic circuits differs from the conventional one in many ways. Firstly, in reversible circuit, only once every output will be taken so that no fan-out should exist. Secondly, every input pattern there should have a distinctive output pattern. In last, an acyclic circuit must be as a result. The reversible gate performs the permutation of only input functions and synthesizes the reversible functions. If a reversible gate contains k inputs and k outputs, then it is addressed as a a $k \times k$ reversible gate. The reversible gates are only included in reversible-based layouts. In reversible designs, the input lines which has constants are known as constant inputs and the outputs which are not used as primary outputs are known as garbage outputs. An optimal design is used to keep minimal number of constant inputs and garbages. Traditional Boolean logic synthesis approaches like Karnaugh Map, Quine–McCluskey, etc. are not allowed to apply directly to synthesize a reversible-based design due to the parameters like Fan-outs, feedback from output gates to input gates are not allowed, number of inputs equal to the numbers of outputs, existence of ancilla inputs and garbage outputs, etc. So implementation only could be possible in the form of cascading of reversible gates.

Classification of synthesis algorithms is shown in Fig. 1; where we have taken the milestone works in this area. All the existing reversible logic synthesis approaches that have been proposed previously can be divided into two major groups: (a) exact approach which produces the optimal solutions, suffered from huge computation time; and (b) heuristic approaches, on the other side which provides near-optimal solutions in short computation time. They can be described in different represen-

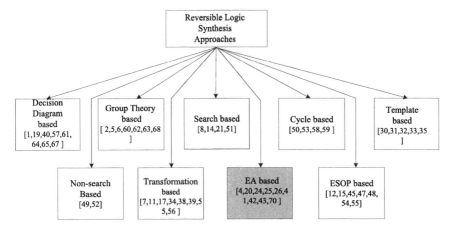

Fig. 1 Classification of synthesis algorithms

tations like BDDs [65], positive polarity Reed–Muller expansion [14], truth-tables [38], matrix representations, permutations [59].

Some of these approaches perform well on smaller designs, but fails when the input count increases, either in requirement of huge computation time and memory, or failure in terms of reaching to the solution. Although having encouraging progress in the field of reversible logic synthesis, still search of best possible synthesis solution remains an open challenge for the researchers. Number of results show the practicability and extend of synthesis are generally not sufficient [25]. So, EA-based algorithm can be used to bolster the potential and efficiency of reversible-based synthesis methods.

In this survey, we present a comparative study of several metaheuristic approaches, algorithms, benchmarks, and future aspects emphasizes to the realization of reversible logic designs. This chapter is organized in following way: Sect. 2 preliminaries are introduced. Section 3 outlines evolutionary approaches. Section 4 includes algorithmic details. Section 5 presents comparison of available benchmarks and discussion which manifest the feasibility of different approaches. Finally, Sect. 6 summarizes this work.

2 GA-Based Synthesis Algorithm

Genetic Algorithm (GA) is based on heuristic search approach and optimization tool for inheriting the method of natural evolution. A GA has the ability for evolving a solution, by exploring the search space with evolutionary heuristics, in same way of genetic information transference known from the Nature [13]. GA related to the broader area of evolutionary algorithms that produces optimal solutions, which are inspired from Darwin's theory of natural evolution, such as fitness function, selection, mutation, and crossover. GA uses an initial population, where each individual within this population is a possible solution for the problem. In GA, an initial population evolves in direction of optimal solutions. GA starts with randomly generated population of individuals and occurs in generations. Each individual of this population evolved in various steps (mutation, crossover or repeated) to produce new generation of individuals for getting a better solution. The fitness function of every individual is evaluated in every generation. Selection considered the fitness function for selecting the individuals from the current population and modified to generate a next population. This new generation is then used in upcoming iterations of the process. Usually, the process terminates after reaching maximum generations or best fitness level has been achieved for the population.

In [70], the authors proposed an array model of reversible- based designs. This paper used universal gates such as wire gate, NOT gate, Toffoli gate and Feynman gates for configuring reversible- based functions. The model consists of cascade of gate-level reversible-based units in a $n \times m$ rectangular array fashion without the feedback and the fan-out as shown in Fig. 2, where n represents maximum input or output wire counts, m represents maximum number of cascade gate-level reversible-based units. Each chromosome generated by the cascade of T types of designable reversible functions and can be encoded by $E = (n \cdot A + n \cdot (p_{max} + q_{max}) \cdot B)$ bits, where $A = \lceil log_2 T \rceil$ bits, $B = \lceil log_2 n \rceil$ bits, p_{max} and q_{max} are the maximum controlled wire bit-count and controlling wire bits for all designable function respectively. New individuals are generated by BSAGA [70]. A pre-bit priority mechanism has been considered to ignore multiplexing error, which happens due to multiplexing between controlling wires and controlled wires.

In literature [27], GA algorithm is used to synthesize an exclusive-or sum-of-product (ESOP)-based structure. This method emphasizes the importance of well-designed encoding method and how it helps in fast convergence of GA. The fitness

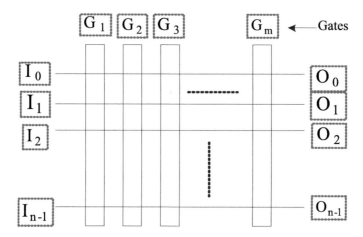

Fig. 2 Array model

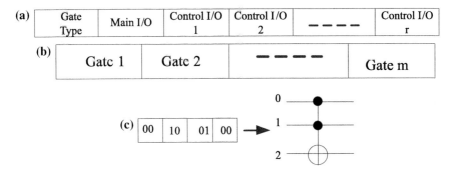

Fig. 3 **a** Coding of a generalized reversible gate. **b** Chromosome of a circuit with m gates. **c** Toffoli gate coding

function can handle any number of inputs and outputs, and suitable for incompletely specified functions. In this work stochastic universal sampling operator is used for selection of individuals because of its population diversity.

In literature [41, 42], aiming at the genetic algorithm (GA) to optimize a given specification. The method also synthesizes incompletely-defined functions. This optimization works more efficiently if the truth table of the given specification contain don't-care conditions and don't-care outputs (garbage outputs). In this chapter, the music line style is used for schematic of design, and a new coding method is proposed to encode a generalized $n \times n$ circuit with m gates, where n is maximum number of parallel input/output lines, and m represents maximum number of columns or gates placed on the parallel lines.

As shown in Fig. 3a some fields are associated with each gate. The first field shows the type of gate (If T types of configurable reversible logic gates available then this field need maximum $\lceil log_2 T \rceil$ bits). The second field indicates the position of its main input or output. The line number of the main output given by this code, in number range from 0 to $n - 1$. The r fields give information about the position of the gate inputs. For instance a 3×3 Toffoli gate shown in Fig. 3c is encoded with 01, 10, 00, 01 (01 is considered as code of Toffoli gate). Figure 3b shows a chromosome, which represents a reversible circuit that contains m gates.

An improved ESOP-based realization of reversible function using genetic algorithm given in [10]. In this Pseudo Kronecker Expressions (PSDKROs) are used for very compact representation and the given algorithm is effective for functions having variables greater than 20.

Manna et al. [28] introduced a searching algorithm based on GA for realizing reversible layouts. This algorithm generates Toffoli gates network for realization of a reversible structure.

Table 1 Comparisons of figure of metrics based on GA

Function name	GA [4]			[70]
	GC	QC	TC	GC
[0, 1, 2, 3, 4, 5, 6,7] 3–17	6	14	56	6
[0, 1, 2, 3, 4, 5, 6, 7] ham3	5	9	48	5
ham7	–	–	–	25
4mod5	5	55	56	5
graycode5	–	–	–	5
graycode6	5	5	40	–
rd32	–	–	–	4
rd53	–	–	–	12
[1, 0, 3, 2, 5, 7, 4, 6] rand_3_9	4	16	48	–
[7, 0, 1, 2, 3, 4, 5, 6] rand_3_1	3	7	24	–
[0, 1, 2, 3, 4, 6, 5, 7] rand_3_2	3	15	48	–
[0, 1, 2, 4, 3, 5, 6, 7] rand_3_3	5	17	64	–
[1, 2, 3, 4, 5, 6, 7, 0] rand_3_4	3	7	24	–
[3, 6, 2, 5, 7, 1, 0, 4] rand_3_5	7	19	72	–
[1, 2, 7, 5, 6, 3, 0, 4] rand_3_6	7	15	64	–
[4, 3, 0, 2, 7, 5, 6, 1] rand_3_7	6	10	48	–
[7, 5, 2, 4, 6, 1, 0, 3] rand_3_8	9	21	80	–
[1, 2, 3, 4, 5, 6, 7, 8, 9, 10, 11, 12, 13, 14, 15, 0] rand_4_3	6	26	64	–
1–bit adder	4	12	48	–

In [4] author's proposed a similar algorithm, which is valid for any gate library of reversible-based logic. So, this chapter encoded generalized Toffoli gates to represent solution. This algorithm is suitable up to 4 or 5 variables and fails to provide solution for larger circuits. To avoid this, factor-based permutation cycles are used. Table 1 shows various functions name and the gate count for each implementation.

3　PSO-Based Synthesis Algorithm

PSO is a stochastically optimization method introduced by Kennedy and Eberhart [18], inspired from the social behavior of creatures such as bird flocking or fish shoaling. In PSO method, particles shows the solution, wander through a multidimensional search space, at every instance each particle modify its position based upon its own experience, and as per the experience of a neighboring particles, maximizing the usage of best positions confronted by itself and its neighboring particles. The method has basic attempt to merge local and global searching techniques in order to detect best feasible solutions.

Table 2 Comparisons of figure of metrics based on PSO

Function name	PSO [3]			[28]
	GC	QC	TC	GC
[1, 0, 3, 2, 5, 7, 4, 6] rand_3_9	5	9	40	4
[7, 0, 1, 2, 3, 4, 5, 6] rand_3_1	3	7	24	3
[0, 1, 2, 3, 4, 6, 5, 7] rand_3_2	3	15	48	3
[0, 1, 2, 4, 3, 5, 6, 7] rand_3_3	5	9	48	5
[1, 2, 3, 4, 5, 6, 7, 0] rand_3_4	3	7	24	4
[3, 6, 2, 5, 7, 1, 0, 4] rand_3_5	7	20	80	7
[1, 2, 7, 5, 6, 3, 0, 4] rand_3_6	6	14	56	7
[4, 3, 0, 2, 7, 5, 6, 1] rand_3_7	6	10	48	6
3_17	6	14	56	–
ham3	5	9	48	–
ex1	4	8	32	–
nth prime	4	8	40	–
miller	5	9	48	–
peres	2	6	24	–
fredkin	3	7	32	–
1bitadder	4	12	48	–
hwb4	12	24	120	–
4mod5(M)	5	55	72	–

Datta et al. [3] presented a PSO-based searching methods to realizing a reversible-based design. The method has ability to find near-optimal solution without searching the complete search space. Each particle in swarm creates a network structure represented as an array of generalized Toffoli gate. The array can be codified as a string of integer [0 to $n \times 2^{n-1}$]. For a reversible design with n line, there are $n \times 2^{n-1}$ possible generalized Toffoli gates. For given specification f this algorithm generates N solutions each of k gates. At the initialization stage, each particle is initialized randomly. Fitness function accounts length of gates, mismatch, hamming distance between present and desire permutation for particles in the swarm. On each iteration the positions of particles changed using well- chosen random function. To accept the new positions for the next iteration, fitness function at new and old position is compared. In [28] similar synthesis algorithm is considered. Fitness function calculated as the ratio of number of matches and the length of the permutation. Table 2 shows various function name and the gate count for each implementation.

4   Ant Colony Optimization-Based Synthesis Algorithm

ACO algorithms are most efficient and widely used algorithm motivated by the forag-
ing behavior of ant colonies [9]. Behavioral patterns exhibited by ants can be explored
to find shortest paths in graphs and similar application. They establish a communication between individuals based on a chemical substance called pheromones, which
they deposit and smell during their search of food from source to nest. This behavior
can be used to develop an algorithm for the solution of optimization problems. The
convergence of ACO depends on pheromone deposit on their forward and backward
travel.

In literature [24], the author's introduced an ACO-based technique for reversible-
based units to formulate the best-path search problem. The approach is capable
to hand massive reversible operations and efficiently produces near-optimal design
or optimal design having less number of gates. A generalized Toffoli gate library
$\{TOF_k | k \leq n\}$ is proposed to implement an n-input reversible-based operation. Each
Toffoli gates is designated as $g(\overrightarrow{c}, t)$, where c represents a vector of control bit out
of n input bits, $\overrightarrow{c} = \{[c_1, c_2, \ldots, c_n] | c_i \varepsilon \{0, 1, 2\}, i = 1, 2, \ldots, n\}$. $c_i = 0$ indicates
positive control bit i ; $c_i = 1$ represents negative control bit i, and $c_i = 2$ shows
neither positive nor negative control line. t represents target bit $t\varepsilon (1, 2, \ldots, n)$. Two
graphs have been proposed. First a probabilistic state transition graph G(V, E) that
correlates the gate selection to help the ants during efficient path search process,
which dynamically updated by pheromone levels and gate count after all the ants
completed their tour. Second a weighted graph G(C, A) called as an ant system
graph (ASGraph) where $C = \{c_1, c_2, \ldots, c_n\}$ is a finite set of elements (reversible
operation). Set comprises of all the arcs (gates) linking the elements. So, the synthesis
of reversible operation able to phrase like a minimization problem on an ASGraph,
in which each arc weight w_{ij} is defined as the lowest cost of gates in a_{ij}.

Key steps in the algorithm:

- *Speculative model for gate selection* (DFS and BFS algorithm are used in local
  search to choose a gate $g(\overrightarrow{c}, t)$. The speculative model decides the target bit and
  the value of each control bit).
- *Initialization of* τ_{pq} (Ant takes decision based on set of pheromone values $\tau_{pq} = (\psi_{pq}, \phi_{pq})$, where τ_{pq} is the amount of pheromone for choosing bit t as target bit,
  and ϕ_{pq} is the amount of pheromone to set control bit i as c_i for target bit t if an
  ant moves from state p to state q. The updated pheromone values are stored in a
  hash table to optimize memory utilization).
- *Pheromone update* (The pheromone graph G(V, E) get updated each time after all
  the ants completed their tour. Gates with less quantum cost get more pheromones
  than other) (Figs. 4a, b and 5).

Sarkar et al. [57] presented a modified version of classical Quine–McCluskey
method under the guidance of ACO techniques has been proposed. Table 3 shows
various function name and the gate count for each implementation.

Fig. 4 **a** State transition graph **b** weighted graph for reversible function

(a)

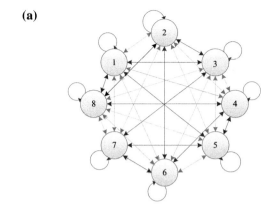

(b)

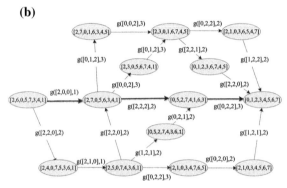

Fig. 5 Realization of reversible function [2, 6, 0, 5, 7, 3, 4, 1] using Toffoli gates

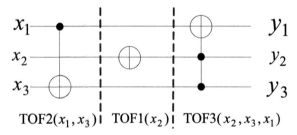

Table 3 Comparisons of figure of metrics based on ACO

Function name	GC	
	ACO based [24]	[57]
[1, 0, 3, 2, 5, 7, 4, 6] rand_3_9	3	5
[7, 0, 1, 2, 3, 4, 5, 6] rand_3_1	3	3
[0, 1, 2, 3, 4, 6, 5, 7] rand_3_2	3	5
[0, 1, 2, 4, 3, 5, 6, 7] rand_3_3	4	6
[0, 1, 2, 3, 4, 5, 6, 8, 7, 9, 10, 11, 12, 13, 14, 15] rand_4_2	7	10
[1, 2, 3, 4, 5, 6, 7, 0] rand_3_4	3	3
[1, 2, 3, 4, 5, 6, 8, 7, 9, 10, 11, 12, 13, 14, 15, 0] rand_4_3	3	4
[0, 7, 6, 9, 4, 11, 10, 13, 8, 15, 14, 1, 12, 3, 2, 5] rand_4_4	4	4
[3, 6, 2, 5, 7, 1, 0, 4] rand_3_5	6	8
[1, 2, 7, 5, 6, 3, 0, 4] rand_3_6	6	7
[4, 3, 0, 2, 7, 5, 6, 1] rand_3_7	5	7
[7, 5, 2, 4, 6, 1, 0, 3] rand_3_8	5	7
[6, 2, 14, 13, 3, 11, 10, 7, 0, 5, 8, 1, 15, 12, 4, 9] rand_4_5	11	14
[2, 9, 7, 13, 10, 4, 2, 14, 3, 0, 12, 6, 8, 15, 11, 1, 5]	11	–
[6, 4, 11, 0, 9, 8, 12, 2, 15, 5, 3, 7, 10, 13, 14, 1]	13	–
[13, 1, 14, 0, 9, 2, 15, 6, 12, 8, 11, 3, 4, 5, 7, 10] rand_4_1	10	–

5 Comparison and Discussion

To evaluate the effectiveness of EA-based synthesis algorithms, we have taken the best results available for each benchmark functions [14, 29, 52]. Table 4 shows the function name and the gate count for each implementation. We compare the best announced algorithm outputs for the NCT library. From Table 4 it is clear that the EA-based algorithm provides better results in terms of gate count (GC) in most of the cases. In other cases, synthesizing circuit using EA results with identical gates. For instance, rand_3_1, rand_3_2, which are implemented with identical gates.

6 Summary

In this paper, a brief algorithmic overview on different types of EA- based synthesis approach has been provided. Comparison with existing work shows EA-based method has better performance in GC, QC, and computational cost. In the last decade, study of reversible circuits and its synthesis methods has received significant attention and the results obtained are quite encouraging, still new efficient synthesis algorithm will remain an open challenge for the researchers.

Table 4 Benchmark comparison

Function name	Gate count					
	MOSAIC	MASLOV	PPRM	GA	PSO	ACO
	[52]	[29]	[14]	[4]	[16]	[24]
[1, 0, 3, 2, 5, 7, 4, 6] rand_3_9	4	–	4	4	5	3
[7, 0, 1, 2, 3, 4, 5, 6] rand_3_1	3	–	3	3	3	3
[0, 1, 2, 3, 4, 6, 5, 7] rand_3_2	3	–	3	3	3	3
[0, 1, 2, 4, 3, 5, 6, 7] rand_3_3	7	–	5	5	5	4
1 bit adder	–	–	–	4	4	4
4mod5	–	4	–	5	5	–
graycode6	–	–	–	5	–	–
graycode5	–	–	–	–	–	–
ham3	–	5	–	3	5	–
3_17	–	6	–	6	6	–
[0, 1, 2, 3, 4, 5, 6, 8, 7, 9, 10, 11, 12, 13, 14, 15] rand_4_2	9	–	7	–	–	7
[1,2,3,4,5,6,7,0] rand_3_4	3	–	3	3	3	3
[1, 2, 3, 4, 5, 6, 8, 7, 9, 10, 11, 12, 13, 14, 15, 0] rand_4_3	4	–	4	–	–	3
[0, 7, 6, 9, 4, 11, 10, 13, 8, 15, 14, 1, 12, 3, 2, 5] rand_4_4	3	–	4	–	–	4
[3, 6, 2, 5, 7, 1, 0, 4] rand_3_5	8	–	7	7	8	6
[1, 2, 7, 5, 6, 3, 0, 4] rand_3_6	8	–	7	7	6	6
[4, 3, 0, 2, 7, 5, 6, 1] rand_3_7	6	–	6	6	6	5
[7, 5, 2, 4, 6, 1, 0, 3] rand_3_8	6	–	7	9	–	5
[6, 2, 14, 13, 3, 11, 10, 7, 0, 5, 8, 1, 15, 12, 4, 9] rand_4_5	19	–	14	–	–	11
[2, 9, 7, 13, 10, 4, 2, 14, 3, 0, 12, 6, 8, 15, 11, 1, 5]	23	–	14	–	–	11
[6, 4, 11, 0, 9, 8, 12, 2, 15, 5, 3, 7, 10, 13, 14, 1]	21	–	17	–	–	13
[13, 1, 14, 0, 9, 2, 15, 6, 12, 8, 11, 3, 4, 5, 7, 10] rand_4_1	29	–	14	–	–	10

References

1. Abdollahi A, Pedram M (2006) Analysis and synthesis of quantum circuits by using quantum decision diagrams. In: Proceedings of design, automation and test in Europe (DATE), vol 1, pp 1–6
2. Al-Rabadi AN (2004) New classes of Kronecker-based reversible decision trees and their group theoretic representations. In: Proceedings of the international workshop on spectral methods and multirate signal processing (SMMSP), pp 233–243
3. Datta K, Sengupta I, Rahaman H (2012) Particle swarm optimization based circuit synthesis of reversible logic. In: Proceedings of international symposium on electronic system design (ISED), pp 226–230
4. Datta K, Sengupta I, Rahaman H (2012) Reversible circuit synthesis using a evolutionary algorithm. In: Proceedings of 25th international conference on computers and devices for communication (CODEC), pp 1–4

5. De Vos A, Van Rentergem Y (2009) Multiple-valued reversible logic circuits. J Mult-Valued Log Soft Comput 15(5/6):489–505
6. De Vos A, Van Rentergem Y, De Keyser K (2006) The decomposition of an arbitrary reversible logic circuit. J Phys A Math Gen 39(18):5015–035
7. Dirac PAM (1939) A new notation for quantum mechanics. Math Proc Camb Philos Soc 144:416–418
8. Donald J, Jha NK (2008) Reversible logic synthesis with fredkin and peres gates. J Emerg Technol Comput Syst 4
9. Dorigo M (1995) Optimization, learning and natural algorithms, PhD thesis, Politecnico di Milano
10. Drechsler R, Finder A, Wille R (2011) Improving ESOP-based synthesis of reversible logic using evolutionary algorithms. In: EvoApplications, vol 2, pp 151–161
11. Dueck GW, Maslov D, Miller DM (2003) Transformation-based synthesis of networks of Toffoli/Fredkin gates. In: Proceedings of the IEEE Canadian conference on electrical and computer engineering, vol 1, pp 211–214
12. Fazel K, Thornton M, Rice JE (2007) ESOP-based Toffoli gate cascade generation. In: Proceedings of the IEEE Pacic Rim conference on communications, computers and signal processing (PACRIM), pp 206–209
13. Goldberg D (1989) Genetic algorithms in search, optimization, and machine learning. Addison-Wesley Professional, Boston
14. Gupta P, Agrawal A, Jha NK (2006) An algorithm for synthesis of reversible logic circuits. IEEE Trans Comput-Aided Des Integr Circuits Syst 25:2317–2330
15. Hamza Z, Dueck GW (2010) Near-optimal ordering of ESOP cubes for Toffoli networks. In: Proceedings of the 2nd annual workshop on reversible computation (RC)
16. Hung WNN, Song X, Yang G, Yang J, Perkowski M (2006) Optimal synthesis of multiple output boolean functions using a set of quantum gates by symbolic reachability analysis. IEEE Trans Comput-Aided Des 25(9):1652–1663
17. Iwama K, Kambayashi Y, Yamashita S (2002) Transformation rules for designing CNOT based quantum circuits. In: Proceedings of the 39th design automation conference, pp 419–424
18. Kennedy J, Eberhart RC (1995) Particle swarm optimization. In: Proceedings of IEEE international conference on neural networks, IEEE Service Center, pp 1942–1948
19. Kerntopf P (2004) A new heuristic algorithm for reversible logic synthesis. In: Proceedings of the design automation conference, pp 834–837
20. Khan MHA, Perkowski M (2004) Genetic algorithm based synthesis of multi-output ternary functions using quantum cascade of generalized ternary gates. In: Proceedings of the 2004 congress on evolutionary computation
21. Kole D, Rahaman H, Das DK, Bhattacharya B (2010) Optimal reversible logic circuits synthesis based on a hybrid dfs-bfs technique. In: International symposium on electronic system design (ISED 2010), pp 208–212
22. Landauer R (1961) Irreversibility and heat generation in the computing process. IBM Res. Dev. 5:183–191
23. Likharev KK (1982) Classical and quantum limitations on energy-consumption in computation. Int J Theor Phys 21:311–325
24. Li M, Zheng Y, Hsiao MS, Huang C (2010) Reversible logic synthesis through and colony optimization. In: Proceedings of design automation test in Europe (DATE 2010), pp 208–212
25. Lukac M, Perkowski M, Gol H (2003) Evolutionary approach to quantum and reversible circuits synthesis. Artif Intell Rev 20(3–4):361–417
26. Lukac M, Perkowski M (2002) Evolving quantum circuits using genetic algorithm. In: Proceedings of the 2002 NASA/DoD conference on evolvable hardware
27. Lukac M, Pivtoraiko M, Mishchenko A, Perkowski M (2002) Automated synthesis of generalized reversible cascades using genetic algorithms. In: Proceedings of the 5th international workshop on Boolean problems, Freiberg, pp 33–45
28. Manna P, Kole DK, Rahaman H, Das DK Bhattacharya BB (2012) Reversible logic circuit synthesis using genetic algorithm and particle swarm optimization. In: Proceedings of international symposium on electronic system design, pp 246–250

29. Maslov D (2005) Reversible logic synthesis benchmarks page. http://www.cs.uvic.ca/ ~dmaslov/
30. Maslov D, Dueck GW, Miller DM (2005) Toffoli network synthesis with templates. IEEE Trans Comput-Aided Des Integr Circuits Syst (TCAD) 24(6):807–817
31. Maslov D, Dueck GW, Miller DM (2003) Fredkin/Toffoli templates for reversible logic synthesis. In: International conference on computer aided design (ICCAD), pp 256–261
32. Maslov D, Dueck GW, Miller DM (2003) Simplification of Toffoli networks via templates. In: Proceedings of the 16th symposium on integrated circuits and system design
33. Maslov D, Dueck GW, Miller DM (2003) Templates for Tffoli network synthesis. In: Proceedings of the international workshop on logic synthesis
34. Maslov D, Dueck GW, Miller DM (2005) Synthesis of Fredkin-Toffoli reversible networks. IEEE Trans Very Large Scale Integr (VLSI) Syst 13(6):765–769
35. Maslov D, Dueck GW, Miller DM (2007) Techniques for the synthesis of reversible Toffoli networks. ACM Trans Des Autom Electron Syst (TODAES) 12(4):42
36. Merkle RC (1993) Two types of mechanical reversible logic. Nanotechnology 4(2):114–131
37. Merkle RC (1993) Reversible electronic logic using switches. Nanotechnology 7:21–40
38. Miller DM, Maslov D, Dueck GW (2003) A transformation based algorithm for reversible logic synthesis. In Proceedings of the design automation conference, pp 318–323
39. Miller DM, Maslov D, Dueck GW (2003) A transformation based algorithm for reversible logic synthesis. In: DAC '03: proceedings of the 40th conference on design automation, pp 318–323. ACM
40. Miller DM, Thornton MA (2006) QMDD: a decision diagram structure for reversible and quantum circuits. In: Proceedings of the IEEE international symposium on multiple-valued logic (ISMVL), p #30 on Proceedings CDROM
41. Mohamadi M, Eshghi M (2008) Heuristic methods to use don't cares in automated design of reversible and quantum logic circuits. Quantum Inf Process J (Springer) 7:175–192
42. Mohammadi M (2012) Efficient genetic based methods for optimizing the reversible and quantum logic circuits. JACR 3(3):85–96
43. Mohammadi M, Eshghi M (2009) On figures of merit in reversible and quantum logic designs. Quantum Inf Process J (Springer) 8(4):297–318
44. Moore GE (1965) Cramming more components onto integrated circuits. Electronics 38(8)
45. Nayeem NM, Rice JE (2011) Improved ESOP-based synthesis of reversible logic. In: Proceedings of the 2011 Reed-Muller workshop, pp 57–62
46. Nielsen M, Chuang I (2000) Quantum computation and quantum information. Cambridge University Press, Cambridge
47. Rice JE, Nayeem N (2011) Ordering techniques for ESOP-based Toffoli cascade generation. In: Proceedings of the IEEE Pacific Rim conference on communications, computers and signal processing (PACRIM), pp 274–279
48. Rice JE, Suen V (2010) Using autocorrelation coefficients-based cost functions in ESOP based Toffoli gate cascade generation. In: Proceedings of the 23rd IEEE Canadian conference on electrical and computer engineering (CCECE), pp 1–6
49. Saeedi M, Sedighi M, Zamani MS (2007) A novel synthesis algorithm for reversible circuits. In: Proceedings of IEEE/ACM International Conference on Computer-Aided Design, pp 65–68
50. Saeedi M, Sedighi M, Zamani MS (2010) A library-based synthesis methodology for reversible logic. Microelectron J 41(4):185–194
51. Saeedi M, Zamani MS, Sedighi M (2007) On the behaviour of substitution-based reversible circuit synthesis algorithms: investigation and improvement. In: Proceedings ISVLSI, pp 428–436
52. Saeedi M, Zamani MS, Sedighi M (2008) Moving forward: a nonsearch based synthesis method toward efficient CNOT-based quantum circuit synthesis algorithms. In: ASPDAC, pp 83–88
53. Saeedi M, Zamani MS, Sedighi M, Sasanian Z (2010) Reversible circuit synthesis using a cycle-based approach. JETC 6(4):13
54. Sanaee Y, Dueck GW (2009) Generating Toffoli networks from ESOP expressions. In: Proceedings of the IEEE Pacific Rim conference on communications, computers and signal processing (PACRIM), pp 715–719

55. Sanaee Y, Dueck GW (2010) ESOP-based Toffoli network generation with transformations. In: Proceedings of the 40th IEEE international symposium on multiple-valued logic (ISMVL), pp 276–281

56. Sanaee Y, Dueck GW (2010) ESOP-based Tooli network generation with transformations. In: Proceedings of the 40th IEEE international symposium on multiple-valued logic (ISMVL), pp 276–281

57. Sarkar M, Ghosal P, Mohanty SP (2013) Reversible circuit synthesis using ACO and SA based Quine-McCluskey method. In: Proceedings of 56th international midwest symposium on circuits and systems (MWSCAS), pp 416–419

58. Sasanian Z, Saeedi M, Sedighi M, Zamani MS (2009) A cycle-based synthesis algorithm for reversible logic. In: Proceedings of Asia South Pacific design automation conference (ASP-DAC), pp 745–750

59. Shende VV, Prasad AK, Markov IL, Hayes JP (2003) Synthesis of reversible logic circuits. IEEE Trans Comput Aided Des 22(6):710–722

60. Storme L, Vos AD, Jacobs G (1999) Group theoretical aspects of reversible logic gates. J Univers Comput Sci 5(5):307–321

61. Thornton M, Miller DM, Goodman D (2006) A decision diagram package for reversible and quantum circuit simulation. In: Proceedings of the IEEE congress on evolutionary computation, pp 8597–8604

62. Van Rentergem Y, De Vos A, Storme L (2005) Implementing an arbitrary reversible logic gate. J Phys A Math Gen 38(16):3555–3577

63. Van Rentergem Y, De Vos A, De Keyser K (2006) Using group theory in reversible computing. In: Proceedings of the 2006 IEEE congress on evolutionary computation (CEC2006), pp 2397–2404

64. Wille Robert, Drechsler Rolf (2010) BDD-based synthesis of reversible logic. Int J Appl Meta-heuristic Comput (IJAMC) 1(4):25–41

65. Wille R, Drechsler R (2009) BDD-Based synthesis of reversible logic for large functions. In: Proceedings of the design automation conference, pp 270–275

66. Wille R, Drechsler R, Oswald C, Garcia-Ortiz A (2012) Automatic design of low power encoders using reversible circuit synthesis. In: Design, automation and test in Europe, pp 1036–1041

67. Wille R, Groe D, Miller DM, Drechsler R (2009) Equivalence checking of reversible circuits. In: Proceedings of the 39th international symposium on multiple-valued logic (IS- MVL), pp 324–330

68. Yang G, Song X, Hung WN, Xie F, Perkowski MA (2006) Group theory based synthesis of binary reversible circuits. Lecture notes in computer science, vol 3959. Springer, pp 365–374

69. Younis SG, Knight TF (1994) Asymptotically zero energy split-level charge recovery logic. In: Workshop low power design, pp 177–182

70. Zhang M, Zhao S, Wang X (2009) Automatic synthesis of reversible logic circuit based on genetic algorithm. In Proceedings of IEEE international conference on intelligent computing and intelligent systems (ICIS 2009), pp 542–546

Part III
Test Approaches

Automatic Error Correction
of Reversible Circuits

M. Fujita

Abstract Automatic correction of logical bugs in reversible circuits is discussed in this chapter. The general logic correction problem can be formulated as Quantified Boolean Formula (QBF) problem which can be solved by repeatedly applying SAT solvers. The automatic correction of reversible circuits can be similarly formulated as QBF. We show various experiments to demonstrate how the method can correct the circuits with an implementation on the logic synthesis and verification tool, ABC from University of California, Berkeley. The discussed methods can be extended to topologically constraints reversible circuit synthesis.

1   Logic Debugging for Reversible Circuits

In this chapter we discuss techniques by which reversible circuits, or more specifically Toffoli networks can be logically debugged. Here we would like to discuss the following situation. Suppose a specification in terms of reversible function is given, and with the method discussed in the previous chapter, or with other techniques, the corresponding reversible circuit is manually generated. Unfortunately, it is shown by simulation and other verification techniques that the reversible circuit is not completely equivalent to the specification, and it must be logically debugged.

Similar situations also happen when the specification is modified after its corresponding circuit is generated. This is so-called "Engineering Change Order" (ECO). When specification changes, instead of regenerating a new circuit from the scratch, it is often better to modify the existing one so that it becomes equivalent to the new specification, since by doing so, the original circuit topology is mostly preserved and the implementation efforts such as layout can also be reused. Logical debugging and ECO are basically doing the same operations, as the goal is to make the given circuit equivalent to the given specification.

Logical debugging basically consists of two steps. The first step is to locate the suspicious portions in the buggy circuit. Typically when a circuit is found to be

M. Fujita (✉)
University of Tokyo, Tokyo, Japan
e-mail: fujita@ee.t.u-tokyo.ac.jp

© Springer Nature Singapore Pte Ltd. 2020
A. K. Singh et al. (eds.), *Design and Testing of Reversible Logic*, Lecture Notes in Electrical Engineering 577, https://doi.org/10.1007/978-981-13-8821-7_8

incorrect by some verification method, some sort of counterexamples are available. By back tracing the dependencies in the simulation results of the counterexamples, which portions in the circuit are in charge of generating wrong values at primary outputs can be examined. If multiple counterexamples are available and buggy portion is to be one portion, intersections of such back traces from the primary outputs on the dependency paths may be most suspicious. Then in the second step, the suspicious portions are actually replaced with new sub-circuits so that the entire circuit becomes equivalent to the specification.

In this chapter, we concentrate on the second step, since the first step is basically the same as the methods for conventional logic circuits. This chapter consists of the following discussions. First, automatic correction techniques which utilize LUT (Look Up Table) based formulation is reviewed with applications to illustrative examples. They are applicable to conventional logic circuits. Then their extensions for reversible circuits are discussed. The key issue is how to generate reversible functions for the LUTs introduced for automatic correction. If the logic functions for the LUTs are generated without the additional constraints shown below, the generated functions are most likely nonreversible. After that, a new formulation of automatic correction of reversible circuits is given using the generalized Toffoli gates discussed later. All of the discussions so far assume that the inputs to the gates or sub-circuits to be corrected do not change. Although this gives simpler methods for the automatic corrections, based on the industrial experiences, around half of the errors cannot be corrected, as they require new inputs to the gates or sub-circuits to be corrected. Therefore, methods which can change the inputs to the gates or sub-circuits to be corrected are also discussed.

Most of the discussions above are made through illustrative examples including the running examples using the logic synthesis and verification tools ABC [1].

2   Automatic Correction Methods Using LUT for Conventional Circuits

There have been proposed a number of techniques on automatic correction of suspicious portions in the buggy circuit. Here one of such techniques based on partial logic synthesis [2, 3] is shown to be useful for reversible circuits as well. This section introduces the partial logic synthesis methods for conventional circuits. It tries to automatically fill the missing portions of the circuit so that the entire circuit becomes equivalent to the specification separately given. For example, suppose that a buggy full adder design is given as the one shown in Fig. 1b and assume that the correct specification is equivalent to the circuit shown in Fig. 1a. Specifically one EOR gate in (a) is wrongly replaced with an AND gate in (b).

There are basically two different formulations on logical debugging or ECO (specification is slightly changes) targeting combinational circuits or combinational parts of sequential circuits:

Fig. 1 The target circuit, full adder

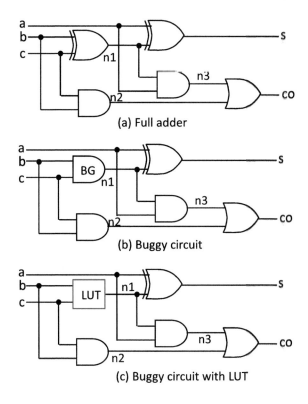

(a) Full adder

(b) Buggy circuit

(c) Buggy circuit with LUT

1. For a set of selected gates, only their types of gates are to be modified in order for the entire circuit to be equivalent to the specification.
2. Same as above, but it is allowed to change not only the types of their gates, but also the fanins (inputs to the gates) can be changed from the original ones to any signals including the primary inputs in the circuit. Please note that as the target is a combinational circuit, by changing the connection, no loop is introduced.

As for problem 1, there have been introduced and researched based on LUT-based problem definition which can be solved by QBF (Quantified Boolean Formula) solvers or incremental SAT solvers [2, 3], and as for problem 2, the problem to find out good replacements for the fanins of the gates is defined as a SAT problem which can be solved by SAT solvers [4].

Here the logic synthesis and ECO methods shown in [2, 3] are briefly introduced through examples, and then the methods shown in [4] is introduced through an example.

Let us consider the circuit shown in Fig. 2. The circuit shown in Fig. 2b is the target buggy circuit to be corrected with respect to the specification shown in the circuit in Fig. 2a. As the first step, the buggy gates whose types are to be modified are identified. This may be accomplished by so-called "path tracing" which traces back the dependencies from the primary outputs whose values are wrong. In this case, the

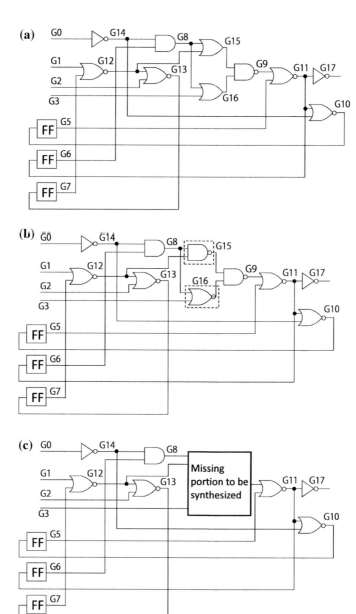

Fig. 2 An example sequential circuit whose combinational part should be debugged

gates, G15 and G16 are identified. Then a single output sub-circuit having those two gates is picked up. The sub-circuit is the region which covers the three gates, G15, G16, and G9. The generated sub-circuit is represented by an LUT having three inputs as shown in Fig. 2c, since there are originally three inputs to the sub circuit. The LUT corresponds to the missing portion in the figure. Please note in this debugging and ECO method, the inputs to the sub-circuit are not allowed to change.

As shown in Fig. 3, the LUT has variables $tt0, \ldots, tt7$ which correspond to the values of the truth table for the sub-circuit and the original three inputs, $in1, in2, in3$. It is essentially equivalent to a multiplexer whose control inputs are $in1, in2, in3$ and the inputs are $tt0, \ldots, tt7$. Please note that the variables $tt0, \ldots, tt7$ show the values of the truth table.

Then the problem to be solved is defined as a QBF (Quantified Boolean Formula) problem:

$\exists tt0, \ldots, tt7. \forall in1, in2, in3.$

Circuit shown in Fig. 2c is equivalent to circuit shown in Fig. 2a.

That is, by assigning appropriate values to $tt0, \ldots, tt7$, for all input values, Circuit shown in Fig. 2c is equivalent to circuit shown in Fig. 2a.

The above problem is not a SAT problem, as both of existential and universal quantifiers appear in the formula. This QBF problem can be efficiently solved by QBF solvers or incremental SAT solvers as shown in [2, 3]. Circuits having tens of thousands of gates can be processed.

Instead of discussing the details of the methods, here they are illustrated with the examples using the logic synthesis and verification tool ABC [1]. In order to describe the gate-level netlists to be processed by ABC, BLIF format is introduced and used here. As for the details of the BLIF format, please see for example, [5].

The circuit shown in Fig. 1a in BLIF format is shown below:

```
1  .model fulladder
2  .inputs a b c
3  .outputs s co
4  .names b c n1
5  10 1
6  01 1
7  .names b c n2
8  11 1
9  .names a n1 n3
10 11 1
11 .names a n1 s
12 10 1
13 01 1
14 .names n3 n2 co
15 1- 1
16 -1 1
17 .end
```

.model in line 1 defines the name of the circuits. .inputs and .outputs in

Fig. 3 3-input LUT (Look
Up Table)

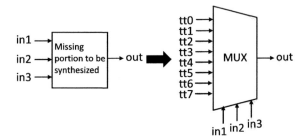

lines 2 and 3 respectively show the list of primary inputs and the list of primary
outputs of the circuit. Then .names defines the functionality of each gate in the
circuit. Variable names following .names are inputs to the gate except for the last
variable. The last variable is the output of the gate. Please note that with .names
statements in BLIF format, all gates must be a single output gate. The lines consisting
of 0, 1, and – after the lines of .names define the logic function of the gates. Here
the input values which make the output of the gate 1 are expressed. That is, the line
11 1 in line 8 means a 2-input AND gate whereas the two lines consisting of 1–
1 and –1 1 in lines 15 and 16 means a 2-input OR gate. – represents a don't care
value, that is, the output does not depend on the value of the variable in this case. In
the same way, the two lines consisting of 10 1 and 01 1 in lines 5–6 and 12–13
means a 2-input EOR gate.

An example of buggy circuit is shown in Fig. 1b. In this example, the buggy gate,
BG is replaced with a 2-input LUT as shown in Fig. 1c. The 2-input LUT in BLIF
format becomes the following:

```
1  .model LUT2
2  .inputs tt0 tt1 tt2 tt3 in1 in2
3  .outputs out
4  .names in1 in2 tt0 tt1 tt2 tt3 out
5  001-- 1
6  01-1- 1
7  10-1- 1
8  11--1 1
9  .end
```

Values of $tt0, \ldots, tt3$ show the values of the truth table of the target 2-input LUT.
Based on the values of $in1$ and $in2$, an appropriate row in the truth table is selected
and connected to *out* which is represented by lines 5–8.

Using the above LUT2 circuit as a sub-circuit, the circuit shown in Fig. 1c is represented in BLIF format as follows:

```
1  .model faLUT
2  .inputs t0 t1 t2 t3 a b c
3  .outputs s co
4  .subckt LUT2 tt0=t0 tt1=t1 tt2=t2 tt3=t3 in1=a in2=b out=n1
5  .names b c n2
```

```
6  11 1
7  .names a n1 n3
8  11 1
9  .names a n1 s
10 10 1
11 01 1
12 .names n3 n2 co
13 1- 1
14 -1 1
15 .end
```

The logic function for the full adder in Sum-of-Products form is shown in Fig. 4a and it can be described in BLIF as below. This becomes the specification for the debugging problem.

```
1  .model fa
2  .inputs a b c
3  .outputs s co
4  .names a b c s
5  001 1
6  010 1
7  100 1
8  111 1
9  .names a b c co
10 -11 1
11 1-1 1
12 11- 1
13 .end
```

In order to solve the debugging problem, a circuit called "miter" (as shown in Fig. 5) is created from the specification in Fig. 4a and the buggy circuit with an LUT in Fig. 4b. The miter circuit in Fig. 5 becomes the followings in BLIF format:

```
1  .model faSyn
2  .inputs t0 t1 t2 t3 a b c
3  .outputs out
4  .subckt faLUT t0=t0 t1=t1 t2=t2 t3=t3 a=a b=b c=c s=s1 co=co1
5  .subckt fa a=a b=b c=c s=s0 co=co0
6  .names s0 s1 out1
7  11 1
8  00 1
9  .names co0 co1 out2
10 11 1
11 00 1
12 .names out1 out2 out
13 11 1
```

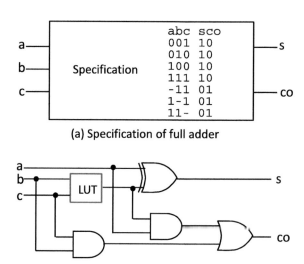

(a) Specification of full adder

(b) Buggy design with LUT

Fig. 4 Specification and buggy circuit with LUT

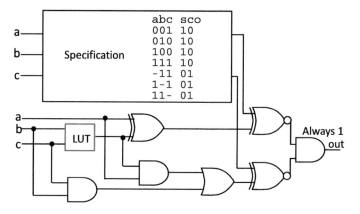

Fig. 5 Miter circuit for the debugging

```
14  .end
```

Then the miter circuit is processed by the ABC command, "qbf". It solves the QBF problem:

$\exists t0, t1, t2, t3. \forall in1, in2, in3.\ faLUT\ circuit\ is\ equivalent\ to\ fa\ circuit.$

Here is the execution trace of the qbf command:

```
abc 06> read faSync.blif
Hierarchy reader flattened 3 instances of logic boxes and left
0 black boxes.
abc 07> qbf -P 4 -v
```

```
Iter 0 : AIG = 0 100 Iter 1 : AIG = 1 010 Syn = 0.00 sec
Iter 2 : AIG = 2 001 Syn = 0.00 sec
Parameters: 0110 Statistics: 0=2 1=5
Solved after 2 iterations. Total runtime = 0.01 sec
```

The parameters: 0110 obtained above are the truth table value and mean that the LUT must represent EOR of the two inputs, which is correct. In the qbf command, the solutions for parameters are iteratively found as seen from the above trace. It tries to find a solution "candidate" by assigning a value set to the input variables, $in1, in2, in3$. First, the values of input variables and truth table variables which make the circuit with LUT nonequivalent to the specification is chosen by a SAT solver. In this case, the values of the input three variables are set to $(1,0,0)$ as seen from the above trace. From this input values, as shown in Fig. 6, one of the parameters is found by a SAT solver such that $t0$ must be 0 in order for the specification and buggy circuit with LUT become equivalent. Then another solution candidate on the values of $t0, t1, t2, t3$ is selected, such as $t0, t1, t2, t3 = (0,0,0,0)$. This solution candidate is falsified by the equivalence checking between the specification and the buggy circuit with LUT programmed as $(0,0,0,0)$. A SAT solver generates a counterexample for the equivalence, which is $(0,1,0)$.

Then as the second iteration, the qbf command tries to find another solution candidate which is correct under both of the two inputs values, $(1,0,0)$ and $(0,1,0)$. As the input values, $(0,1,0)$ forces $t1$ to be 1 as shown in Fig. 7, a solution candidate, such as $t0, t1, t2, t3 = (0,1,0,0)$ is generated. Again a counterexample for this solution candidate is the input values, $(0,0,1)$. Now as the third iteration, the qbf command tries to find another solution candidate which is correct under all of the three inputs values, $(1,0,0)$, $(0,1,0)$, and $(0,0,1)$. In this case, the qbf command find an actually correct solution of $t0, t1, t2, t3 = (0,1,1,0)$, and the qbf command terminates, as there is no more counter example.

As can be seen from the above, three input values are needed instead of 8 possible value combinations. Generally speaking, the algorithm used in qbf command finds a correct solution with a small number of iterations in most of the cases as seen in [2, 3].

3   LUT Based Automatic Correction for Reversible Circuits

The method presented in the last section can be directly applied to reversible circuits under the context shown in Fig. 8. A Toffoli network is given as shown in the top of the figure. If it is logically buggy, the target Toffoli gate is identified just like the discussions in the previous section. Then it is replaced with an LUT by which the entire Toffoli network becomes correct. Then the functionality of the LUT can be determined by the same way as the method in [2, 3]. The functionality realized by the Toffoli gates other than the buggy one can be translated into regular logic formulae which become the constraints on the problem above. This translation is

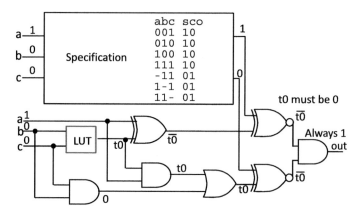

Fig. 6 Miter circuit for the debugging

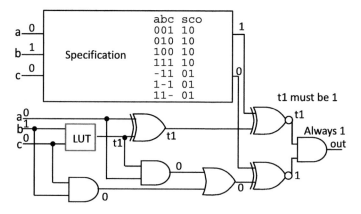

Fig. 7 Miter circuit for the debugging

rather straightforward, as what must be performed is simply a manipulation of logic formulae. One difference is that as Toffoli networks have a number of lines, when finding corrections, up to the number of lines of LUTs may be required for the problem definition, which is within the methods shown in [2, 3] only if the number of lines is not so large (around 16 or less). This is because in order to define LUTs, exponential numbers of variables for rows of the truth tables are required.

There is, however, one problem. As the circuit must be reversible and implemented as (potentially cascades of) Toffoli gate, the new logic function obtained for the LUT as the solution of the QBF problem should also be a reversible function. For that first of all, the number of inputs and the number of outputs must be the same. Also when generating the solution for the QBF problem, the function generated must be a reversible function. This can be checked by the following way for functions for LUTs.

Fig. 8 Debugging Toffoli circuits

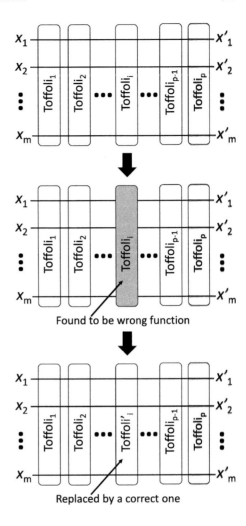

Found to be wrong function

Replaced by a correct one

A reversible function is basically permuting input values and generate them. The input value and the output value must be a one-to-one mapping. This means that in order to be a reversible function, the output values can become the same only if the input values are the same. As long as the input values are different, they must generate different output values. On the contrary, if this is satisfied, all of the input values must be realized at the output, and so the function is reversible.

Theorem 1 *Given a n inputs and n outputs LUT, let the n functions defined by the LUT be $LUT_1(in_1, in_2, \ldots, in_n)$, $LUT_2(in_1, in_2, \ldots, in_n)$, $\ldots$, $LUT_n(in_1, in_2, \ldots, in_n)$. Also let $(in_{11}, in_{12}, \ldots, in_{1n})$ and $(in_{21}, in_{22}, \ldots, in_{2n})$ be two input values. The overall n inputs and n outputs function realized by the LUT is reversible if and only if the disjunction of the following two is always true:*
(1) $((in_{11} = in_{21}) \wedge (in_{12} = in_{22}) \wedge \cdots \wedge (in_{1n} = in_{2n}))$

(2) For some i $(1 \le i \le n)$. $(LUT_i(in_{11}, in_{12}, \ldots, in_{1n}) \ne LUT_i(in_{21}, in_{22}, \ldots, in_{2n}))$

The condition (2) above consists of n inequalities. So the solution for the debugging problem with LUT-based formulation must also satisfy the theorem. The condition for the theorem in the case of three inputs can be described in BLIF format as follows:

```
1  .model RevCheck
2  .inputs T00 T01 T02 T03 T04 T05 T06 T07 T10 T11 T12 T13 T14
T15 T16 T17 T20 T21 T22 T23 T24 T25 T26 T27 in11 in12 in13
in21 in22 in23
3  .outputs out
4  .subckt LUT33 t00 = T00 t01 = T01 t02 = T02 t03 = T03 t04 =
T04 t05 = T05 t06 = T06 t07 = T07 t10 = T10 t11 = T11 t12 =
T12 t13 = T13 t14 = T14 t15 = T15 t16 = T16 t17 = T17 t20 =
T20 t21 = T21 t22 = T22 t23 = T23 t24 = T24 t25 = T25 t26 =
T26 t27 = T27 in1=in11 in2=in12 in3=in13 out1=o11 out2=o12
out3=o13
5  .subckt LUT33 t00 = T00 t01 = T01 t02 = T02 t03 = T03 t04 =
T04 t05 = T05 t06 = T06 t07 = T07 t10 = T10 t11 = T11 t12 =
T12 t13 = T13 t14 = T14 t15 = T15 t16 = T16 t17 = T17 t20 =
T20 t21 = T21 t22 = T22 t23 = T23 t24 = T24 t25 = T25 t26 =
T26 t27 = T27 in1=in21 in2=in22 in3=in23 out1=o21 out2=o22
out3=o23
6  .names in11 in21 y1
7  11 1
8  00 1
9  .names in12 in22 y2
10 11 1
11 00 1
12 .names in13 in23 y3
13 11 1
14 00 1
15 .names y1 y2 y3 y
16 111 1
17 .names o11 o21 z1
18 11 1
19 00 1
20 .names o12 o22 z2
21 11 1
22 00 1
23 .names o13 o23 z3
24 11 1
25 00 1
26 .names z1 z2 z3 z
```

```
27 111 1
28 .names y z out
29 1- 1
30 -0 1
31 .end
```

The following way of applying the qbf command finds a reversible function.

```
abc 02> qbf -P 24
Parameters: 110010011110010010010101 Statistics: 0=12 1=18
Solved after 25 iterations. Total runtime = 0.08 sec
```

Please note that the above execution has over 25 iterations, as in total 24 variables are existentially quantified. By combining the constraints for a function to be reversible, shown above with the constraints for LUT-based debugging, solutions for the LUT which are reversible functions can be obtained.

This method, however, may not scale well if the number of inputs to LUTs becomes large, as the constraints for a function to be reversible needs a number of variables to be defined. So instead of doing above, a debugging method with the generalized Toffoli gate is introduced in the next section.

4 Debugging Based on Universal Toffoli Gates

An alternative approach is to represent the functionality of possible Toffoli gates with a set of parameter variables instead of using LUT. Here an example of a debugging process is briefly shown with ABC tool using this idea. The target circuits are shown

Fig. 9 Example of debugging a Toffoli circuit

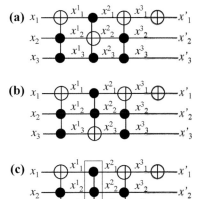

in Fig. 9. (a) is assumed to be a correct circuit whereas (b) is a buggy one as can be seen in the second Toffoli gate in (b).

The first step is to represent the general functionality of a Toffoli gate with a set of parameter variables. Two sets of parameter variables are introduced. The first set is to show whether each line in the Toffoli network is connected to the control input of the Toffoli gate or not. If m_i, $1 \le i \le 3$ is 1, then line_i is one of the control inputs of the Toffoli gate. Clearly at least one of the lines must not be connected to the control input of the Toffoli gate, and so $\overline{m_1 \wedge m_2 \wedge m_3}$ must be satisfied (must be always 1).

The second set of parameter variables is to show whether each line is the target of the Toffoli gate or not. If e_i is 1, line_i is the target line. Therefore, for example, the following holds:

$$x_1^1 = (\overline{e_1} \wedge x_1) \vee (e_1 \wedge (((\overline{m_1} \vee x_1) \wedge (\overline{m_2} \vee x_2) \wedge (\overline{m_3} \vee x_3)) \rightarrow (((\overline{m_1} \vee x_1)$$
$$\wedge (\overline{m_2} \vee x_2) \wedge (\overline{m_3} \vee x_3)) \oplus x_1)))$$

If e_1 is 0, that is line_1 is not the target line, x_1^1 should have the same value of x_1, i.e., $x_1^1 = x_1$ (or $\overline{e_1} \wedge x_1$). If e_1 is 1, line_1 is the target line, and so the value of x_1^1 should be:

$$((\overline{m_1} \vee x_1) \wedge (\overline{m_2} \vee x_2) \wedge (\overline{m_3} \vee x_3)) \oplus x_1).$$

Finally the same line cannot be both of control input and the target of the Toffoli gate: $\overline{m_1} \vee \overline{e_1}$.

The above is just for line_1, and there must be corresponding constraints for line_2 and line_3 as well.

Also as exactly one of the lines must be a target of the Toffoli gate, the following must be satisfied: $(m_1 \vee m_2 \vee m_3) \wedge (\overline{m_1 \wedge m_2} \wedge \overline{m_2 \wedge m_3} \wedge \overline{m_3 \wedge m_1})$. Clearly a line cannot become both of the control input and target line at the same time:

$$\overline{e_1 \wedge m_1}$$
$$\overline{e_2 \wedge m_2}$$
$$\overline{e_3 \wedge m_3}$$

In the case of three lines, these are summarized as follows:

$$O_1 \equiv \overline{e_1 \wedge m_1}$$
$$O_2 \equiv \overline{e_2 \wedge m_2}$$
$$O_3 \equiv \overline{e_3 \wedge m_3}$$
$$P \equiv (\overline{m_1} \vee x_1) \wedge (\overline{m_2} \vee x_2) \wedge (\overline{m_3} \vee x_3)$$
$$Q \equiv \overline{m_1 \wedge m_2 \wedge m_3}$$
$$R_0 \equiv e_1 \vee e_2 \vee e_3$$
$$R_1 \equiv \overline{e_1 \wedge e_2}$$
$$R_2 \equiv \overline{e_2 \wedge e_3}$$
$$R_3 \equiv \overline{e_2 \wedge e_1}$$
$$S_1 \equiv \overline{e_1} \wedge x_1$$
$$S_2 \equiv \overline{e_2} \wedge x_2$$
$$S_3 \equiv \overline{e_3} \wedge x_3$$
$$T_1 \equiv e_1 \rightarrow (P \oplus x_1)$$
$$T_2 \equiv e_2 \rightarrow (P \oplus x_2)$$
$$T_3 \equiv e_3 \rightarrow (P \oplus x_3)$$
$$x_1^1 \equiv \text{Given by the specification}$$

$x_2^1 \equiv$ Given by the specification

$x_3^1 \equiv$ Given by the specification

$N_1 \equiv (x_1^1 = S_1 \vee (e_1 \wedge T_1))$

$N_2 \equiv (x_2^1 = S_2 \vee (e_2 \wedge T_2))$

$N_3 \equiv (x_3^1 = S_3 \vee (e_3 \wedge T_3))$

Then overall constraints become:

$O_1 \wedge O_2 \wedge O_3 \wedge Q \wedge R_0 \wedge R_1 \wedge R_2 \wedge R_3 \wedge N_1 \wedge N_2 \wedge N_3$

The values of x_1^1, x_2^1, x_3^1 represent the output values of the three lines.

The above formulae can be specified in ABC as a combinational circuit with one primary output in BLIF format. The constraints mean the primary output of the circuit must be always 1. When writing a combinational circuit for the constraints, input variables are defined in such a way that parameter variables ($m_1, m_2, m_3, e_1, e_2, e_3$) appearing first followed by the primary input variables (x_1, x_2, x_3). Once that is done, the qbf command in ABC can solve the problem.

Please note that the above way of defining SAT problem is a simplified version of the method discussed in the exact synthesis approach of the previous chapter [6–8].

A BLIF file of the above constraints assuming that there are three output lines should realize:

$x_1^1 = x_1 \oplus x_2, x_2^1 = x_2, x_3^1 = x_3.$

This can be made sure by seeing the lines 69–71 for x_1^1, lines 75–76 for x_2^1, and lines 80–81 for x_3^1.

```
1  .model exreversible
2  .inputs m1 m2 m3 e1 e2 e3 x1 x2 x3
3  .outputs out
4  .names e1 m1 O1
5  11 0
6  .names e2 m2 O2
7  11 0
8  .names e3 m3 O3
9  11 0
10 .names m1 x1 P1
11 0- 1
12 -1 1
13 .names m2 x2 P2
14 0- 1
15 -1 1
16 .names m3 x3 P3
17 0- 1
18 -1 1
19 .names P1 P2 P3 P
20 111 1
21 .names m1 m2 m3 Q
22 111 0
23 .names e1 e2 e3 R0
24 1- 1
```

```
25 -1- 1
26 -1 1
27 .names e1 e2 R1
28 11 0
29 .names e2 e3 R2
30 11 0
31 .names e3 e1 R3
32 11 0
33 .names e1 x1 S1
34 01 1
35 .names e2 x2 S2
36 01 1
37 .names e3 x3 S3
38 01 1
39 .names P x1 T11
40 10 1
41 01 1
42 .names P x2 T12
43 10 1
44 01 1
45 .names P x3 T13
46 10 1
47 01 1
48 .names e1 T11 T1
49 0- 1
50 -1 1
51 .names e2 T12 T2
52 0- 1
53 -1 1
54 .names e3 T13 T3
55 0- 1
56 -1 1
57 .names S1 T1 e1 ST1
58 1- 1
59 -11 1
60 .names S2 T2 e2 ST2
61 1- 1
62 -11 1
63 .names S3 T3 e3 ST3
64 1- 1
65 -11 1
66 .names x11 ST1 N1
67 11 1
68 00 1
69 .names x1 x2 x11
```

```
70 10 1
71 01 1
72 .names x12 ST2 N2
73 11 1
74 00 1
75 .names x2 x12
76 1 1
77 .names x13 ST3 N3
78 11 1
79 00 1
80 .names x3 x13
81 1 1
82 .names O1 O2 O3 Q R0 R1 R2 R3 N1 N2 N3 out
83 11111111111 1
84 .end
```

The BLIF file is read by ABC tool and then qbf command is executed in ABC as follows:

```
abc 02> qbf -P 6 -v
Iter 0 : AIG = 27 110
Iter 1 : AIG = 38 101 Syn = 0.01 sec
Parameters: 010100 Statistics: 0=4 1=5
Solved after 1 interations. Total runtime = 0.02 sec
```

The execution of qbf command generates the values for the parameters. 010100. As the order of inputs in Fig. 9 is m_1, m_2, m_3, e_1, e_2, e_3, this indicates the following:
$m_1 = 0, m_2 = 1, m_3 = 0, e_1 = 1, e_2 = 0, e_3 = 0$
With these parameter values, the functionality of the three lines, x_1^1, x_2^1, x_3^1 becomes the following as intended.
$x_1^1 = x_1 \oplus x_2, x_2^1 = x_2, x_3^1 = x_3$

For the debugging of the entire circuit shown in Fig. 9c, the above generalized Toffoli gate is cascaded by four times. The first, third, and fourth generalized Toffoli gates are programmed following the functionality shown in Fig. 9c as follows:
$x_1^1 = x_2 \wedge x_3 \oplus x_1, x_2^1 = x_2, x_3^1 = x_3$
$x_1^3 = x_2^2 \wedge x_3^2 \oplus x_1^2, x_2^3 = x_2^2, x_3^3 = x_3^2$
$x_1^4 = \overline{x_1^3}, x_2^4 = x_2^3, x_3^4 = x_3^3$
The second generalized Toffoli gate is the target to be corrected, and so its parameter values are to be synthesized by the qbf command.

For example, in the first Toffoli gate in Fig. 9c, the target line is line$_1$ and the lines, line$_2$ and line$_3$ are control lines. This means $m_1 = 0, m_2 = 1, m_3 = 1, e_1 = 1, e_2 = 0, e_3 = 0$ and can be described in BLIF format as:

```
.subckt Toffoli m1=zero m2=one m3=one e1=one e2=zero
e3=zero in1=li1 in2=li2 in3=li3 out1=x11T out2=x12T
out3=x13T out=invar0
```

where zero and one are defined as in BLIF format:

```
.names zero
.names zero one
0 1
```

The third and fourth gates can be described in the same way. By combining these, the buggy circuit in Fig. 9c can be efficiently corrected with respect to the specification in (a).

The overall constraints for the debugging of Fig. 9c with respect to the specification in (a) becomes the following in BLIF format:

```
1  .model Toffolinetwork
2  .inputs M1 M2 M3 E1 E2 E3 Li1 Li2 Li3
3  .outputs out
4  .names zero
5  .names zero one
6  0 1
7  .names lo1 Slo1 ooo1
8  11 1
9  00 1
10 .names lo2 Slo2 ooo2
11 11 1
12 00 1
13 .names lo3 Slo3 ooo3
14 11 1
15 00 1
16 .names ooo1 ooo2 ooo3 out
17 111 1
18 .subckt Spec li1=Li1 li2=Li2 li3=Li3 lo1=Slo1
lo2=Slo2 lo3=Slo3
19 .subckt Toffoli m1=zero m2=one m3=one e1=one
e2=zero e3=zero in1=Li1 in2=Li2 in3=Li3
out1=x11T out2=x12T out3=x13T out=invar0
20 .names x11T invar0 x11
21 11 1
22 .names x12T invar0 x12
23 11 1
24 .names x13T invar0 x13
25 11 1
26 .subckt Toffoli m1=M1 m2=M2 m3=M3 e1=E1 e2=E2 e3=E3
in1=x11 in2=x12 in3=x13 out1=x21T out2=x22T out3=x23T
out=invar1
27 .names x21T invar1 x21
28 11 1
29 .names x22T invar1 x22
30 11 1
```

```
31 .names x23T invar1 x23
32 11 1
33 .subckt Toffoli m1=zero m2=one m3=one e1=one e2=zero
e3=zero in1=x21 in2=x22 in3=x23 out1=x31T out2=x32T
out3=x33T out=invar2
34 .names x31T invar2 x31
35 11 1
36 .names x32T invar2 x32
37 11 1
38 .names x33T invar2 x33
39 11 1
40 .subckt Toffoli m1=zero m2=zero m3=zero e1=one e2=zero
e3=zero in1=x31 in2=x32 in3=x33 out1=x41T out2=x42T
out3=x43T out=invar3
41 .names x41T invar3 x41
42 11 1
43 .names x42T invar3 x42
44 11 1
45 .names x43T invar3 x43
46 11 1
47 .names x41 lo1
48 1 1
49 .names x42 lo2
50 1 1
51 .names x43 lo3
52 1 1
53 .end
```

The automatic correction can be performed by the qbf command of ABC just like before:

```
abc 02> qbf -P 6 -v
Iter 0 : AIG = 34 000
Iter 1 : AIG = 52 010 Syn = 0.01 sec
Iter 2 : AIG = 69 110 Syn = 0.00 sec
Iter 3 : AIG = 87 100 Syn = 0.00 sec
Parameters: 110001 Statistics: 0=3 1=6
Solved after 3 interations. Total runtime = 0.02 sec
```

From the above the solution for the parameter variables are

$m_1 = 1, m_2 = 1, m_3 = 0, e_1 = 0, e_2 = 0, e_3 = 1$

which is the same as the circuit in Fig. 9a.

Fig. 10 Bug example: need to change both inputs and the function of the gate

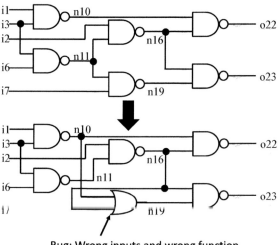

Bug: Wrong inputs and wrong function

5   Fanin

As for the second problem, that is not only the types of gates but also inputs (fanin) to the gates must be changed, the method shown in [4] can be directly applied before applying the methods shown in [2, 3]. The method in [4] can find all possible sets of fanins by which the entire circuit can become correct. For example, the top circuit in Fig. 10 is the correct one, and the bottom circuit is an incorrect buggy one. As can be seen from the figure, the inputs and type of the gate, n19 is wrong. The search problem for the fanin of the gate can be formulated as a SAT and set-covering problem [4], and large circuits having hundreds of thousands of gates can be analyzed with the method in [4]. Moreover, various costs can be defined, and the fanins with minimum cost can be searched. Essentially search problem becomes a set-covering problem.

Normally the lowest cost solution is selected, but for reversible circuits, other criteria can be defined for the optimization of the entire circuit. As discussed above, the hierarchical approach is promising especially for larger circuits. With the method for the fanin selection, LUT networks can be optimized or restructured so that such hierarchical approach works better. For example, reducing the numbers of fanin of LUTs is generally good even for reversible circuits, but that may depend on the functions that must be realized by the LUTs. This is one of the interesting future research topics.

6   Conclusions

In this chapter, debugging techniques for reversible circuits which are extensions over the debugging methods for conventional circuits have been presented by showing sample execution traces with the logic synthesis and verification tool, ABC. The above BLIF descriptions can be directly processed by ABC with the qbf command for debugging.

Although the presented methods work for medium sizes of reversible circuits, they need to be extended in order to deal with large circuits. In order to realize that, various heuristics must be developed on the selections of lines which are inputs to Toffoli gates. For that, the fanin selection techniques briefly discussed in the previous section can be a starting point. Also, how the methods presented in this chapter can be combined with hierarchical approaches [9, 10] is to be explored.

References

1. Brayton RK, Mishchenko A (2010) ABC: an academic industrial-strength verification tool. In: 22nd International Conference on Computer Aided Verification (CAV 2010), pp 24–40
2. Fujita M, Jo S, Ono S, Matsumoto T (2013) Partial synthesis through sampling with and without specification. In: ICCAD
3. Fujita M (2015) Toward unification of synthesis and verification in topologically constrained logic design. Proc. IEEE 103(11):2052–2060
4. Gharehbaghi AM, Fujita M (2017) A new approach for selecting inputs of logic functions during debug. In: International Symposium on Quality Electronic Design (ISQED 2019)
5. Sentovich EM, Singh KJ, Moon C, Savoj H, Brayton RK, Sangiovanni-Vincentelli A (1992) Sequential circuit design using synthesis and optimization. In: IEEE International Conference on Computer Design: VLSI in Computers & Processors
6. Shende VV, Prasad AK, Markov IL, Hayes JP (2002) Reversible logic circuit synthesis. In: ACM/IEEE International Conference on Computer Aided Design (ICCAD), San Jose, USA, pp 125–132
7. Große D, Wille R, Dueck GW, Drechsler R (2009) Exact multiple control Toffoli network synthesis with SAT techniques. IEEE Trans CAD 28(5):703–715
8. Wille R, Soeken M, Przigoda N, Drechsler R (2012) Exact synthesis of Toffoli gate circuits with negative control lines. In: IEEE 42nd International Symposium on Multiple-Valued Logic, Victoria, Canada, pp 69–74
9. Soeken M, Chattopadhyay A (2016) Unlocking efficiency and scalability of reversible logic synthesis using conventional logic synthesis. In: ACM/IEEE Design Automation Conference
10. Soeken M, Roetteler M, Wiebe N, De Micheli G (2017) Hierarchical reversible logic synthesis using LUTs. In: ACM/IEEE Design Automation Confernece, Austin, USA

Fault Models and Test Approaches in Reversible Logic Circuits

H. M. Gaur, T. N. Sasamal, A. K. Singh and A. Mohan

Abstract The operations in reversible circuits are fully controllable and observable due to their bijective property which provides cursive testing. Testing can be categorized into two behavioral schemes: (i) Online testing, where the detection of faults within the circuit is carried out during its operation, (ii) Offline testing, where test vectors are applied after extracting the circuit out from its normal operation and the correct output values are known. Test data minimization for a specific kind of fault model such as stuck-at, bridging, missing gate, cross-point, and cell faults, is an important factor in this type of testing using meta-heuristic algorithms and circuit modification methodologies. Diverse varieties of fault families in reversible logic circuits and an exclusive study of testable design advances for these faults are portrayed in this chapter. Plentiful approaches were projected under two extensive classifications to meet the challenge. The methodologies are alleged to coat almost all the faults and their sub kind by exploiting the properties of reversible gates and circuits. The objective is to minimize testing overhead, which can be achieved by reducing the cost metrics utilized for testability.

H. M. Gaur (✉)
Department of Electronics & Communication Engineering, ABES Institute of Technology,
Ghaziabad (Delhi NCR) 201009, India
e-mail: leoharimohan84@gmail.com

T. N. Sasamal
Department of Electronics and Communication Engineering,
NIT, Kurukshetra, India
e-mail: sasamal.trailokyanath@gmail.com

A. K. Singh
Department of Computer Applications, NIT, Kurukshetra, India
e-mail: ashutosh@nitkkr.ac.in

A. Mohan
Department of Electronics Engineering, IIT-BHU, Varanasi, India
e-mail: profanandmohan@gmail.com

© Springer Nature Singapore Pte Ltd. 2020
A. K. Singh et al. (eds.), *Design and Testing of Reversible Logic*, Lecture Notes
in Electrical Engineering 577, https://doi.org/10.1007/978-981-13-8821-7_9

1 Introduction

In quantum circuits, de-coherence enforces the quantum states of bits to decay. It results in loss of information that causes faults. Due to this phenomenon, these circuits are more prone to faults than conventional circuits [1]. As reversible circuits have direct relations to quantum circuits, they are largely prone to several transient and permanent faults which cause single and multiple point failures. Testing assures the correct operations of logic circuits which show its untamed necessity and their excellence. The operations in reversible circuits are fully controllable and observable due to their bijective property which provides cursive testing. It has also been extensively studied since the last decade by exploiting this property for the identification of several types of fault models. These fault models are stuck-at, bridging, missing gate, cross point, and cell faults. Testing can be categorized in two behavioral schemes First is online testing, where the detection of faults within the circuit is carried out during its operation [2, 3]. It provides a built-in self- testable environment through design methodology and circuit modification for the detection of fault models in terms of single and multiple-bit-flip faults. Second is offline testing, where test vectors are applied after extracting the circuit out from its normal operation and the correct output values are known. Test data minimization for a specific kind of fault model is an important factor in this type of testing using meta-heuristic algorithms and circuit modification methodologies.

Several novel paradigms have been presented in both the area of testing reversible logic circuits. Built-in testable environment are provided over pristine fundamental (MCT and MCF) and new gates-based design methodologies [4–11] and modification principles [12–16]. Test data minimization is achieved over new deterministic [17–27], randomized test pattern generation algorithms [28–31] and modification techniques [32–37] for respective faults. It is noticed that, the technique of parity checking [38] is dominating for the recognition of single/multiple bit faults in online testing, whereas the bijectivity property of reversible logic circuits is utilized in case of offline testing. The reduction of operating cost has been achieved to some needful extent in all the proposed approaches with respect to prior ones to narrow the compensation with overall testing overheads. Fault tolerance is also the architectural attribute of a digital system that maintains proper functioning of a logic machine during various kinds of failures [39, 40]. The inclusion of fault tolerance abilities during the design and synthesis process is also in the development phase [41–46].

Testing impetus a dramatic increment in the utilization of resources in terms of cost and power requirements. In electronic circuits, it leads to a large increment in operating costs from their original circuits like gates and wires, which drastically increases size and power. It accounts 30–60% of the cost of manufacturing electronic devices by consuming extra hardware and resources [47]. During the construction of built-in testable reversible circuits over design methodology or modification, the operating cost increases in terms of gates, wires, ancilla input, and garbage output.

Test data minimization over meta-heuristic algorithms consume excess time and a separate hardware to produce test sets. Test data minimization over modification includes incorporation of additional hardware as well as time to run test vectors for the respective type of fault. Hence, necessary actions should be taken to reduce excess usage of hardware and time to lower the overall testing overheads and narrow the compensation with power.

2 Fault Models

Faults are a type of deficiency in a circuit which reflects in the imperfect behavior and functional abilities of a system long lastingly or for a limited period of time. They can be occurred due to any human and environmental issues [2] and termed as permanent and nonpermanent faults respectively. A fault model depicts the category of fault occurrence in a circuit and guides in identifying the target for testing. Numerous fault models are projected in the past along with respective categorization in reversible circuits as labeled out in Fig. 1. Following is the short description of these fault models given in this figure.

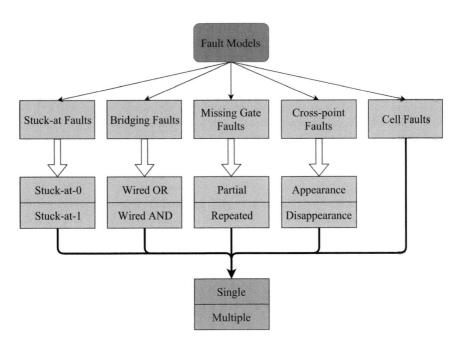

Fig. 1 Faults in reversible circuits

Fig. 2 Stuck-at and bridging faults

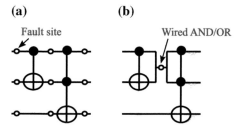

Stuck-At Fault

Alike conventional logic circuits, this fault occurred when any wire in a circuit get settled on a single logic bits 0 or 1 are termed as stuck-at-0 (S-a-0) or stuck-at-1 (S-a-1) faults respectively. The faults can be aroused on single or multiple nodes at a single time which can be either of same or of both types. In reversible circuits, these faults occurred at the input of the gates and the input/output wires of the circuit. The total number of stuck-at faults (S-a-0 and S-a-1) in the circuit is given by $\{2 \times (\sum_{i=1}^{G} g_i + n)\}$ where, G are the number of gates contained by the n wires circuit and g_i is the gate size of ith gate. For example, there are nine possible sites for this type of fault occurrence are shown by tiny circles in Fig. 2a.

Bridging Fault

This category of fault arises when two or more adjacent wires of a circuit get physically come in contact or linked and resemble the abilities of wired AND or OR interconnections that consequence into erroneous functionality. The illustration is provided in Fig. 2b which, these faults may occur on a couple of wires or on multiple wires at respective levels of the circuit which are termed as single intra-level and multiple intra-level bridging faults. The number of levels of the circuit is governed by number of gates. For instance, a circuit containing G gates there will be $G + 1$ distinct levels which include the input of each gate and the final output. The number of single intra-level faults between a couple of wires are given by $^{n}C_2$ and all single intra-level faults between a couple of wires in the circuit is given by $\{(G + 1) \times ^{n}C_2\}$. The circuit represented in Fig. 2b has three levels and number of single intra-level faults between a couple of wires are 9.

Missing Gate Fault

When a gate in a circuit fully fails to perform its characteristics or act completely disappeared from the circuit is called a single missing-gate fault (SMGF), as illustrated in Fig. 3a. As a result, the output changes to a faulty value as illustrated in the figure where the fault-free/faulty logic values are written on every level of the circuit.

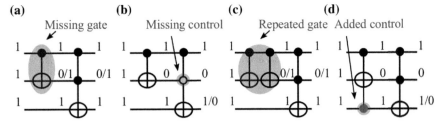

Fig. 3 Missing gate and cross-point faults

Multiple missing gate fault (MMGF) is the disappearance of two or more successive gates. Maximum occurrence of SMGF in the circuit is equal to number of gates (G) contained by the circuit and number of MMGF is given by $\{G(G + 1)\backslash 2 - G\}$. Hence, the total number MGF (SMGF and MMGF) in the circuit can be calculated by $\{G(G + 1)\backslash 2\}$. For instance, number of MGF in the circuit for Fig. 3a is 3.

The of missing gate faults are also labeled as partial missing gate fault (PMGF) and repeated gate fault (RGF) by the researchers of this area. PMGF occurred when any control point is disappeared from a gate and RGF is a replication of operation by a single gate for multiple instances. The illustrations of these faults can be acknowledged in Fig. 3b and c respectively. There is no effect of RGF on the circuit if the instances are even in number. For odd number of instances, RGF will result as that of SMGF in the circuit.

Cross-Point Fault

This fault is associated to the nonfunctioning and inclusion of control points of reversible gates in a circuit. These faults are referred to as appearance (AF) and disappearance (DF) types of faults in the circuit. The behavior of these faults can be acknowledged from the Figs. 3d and b respectively. Disappearance faults show similar tendencies as that of PMGFs, the difference seems in the names only according to the researchers from the past. Total number of single AF in the circuit can be calculated by $(nG - \sum_{i=1}^{G} g_i)$ and number of single DF can be given by $(\sum_{i=1}^{G} g_i - G)$.

Cell Fault

Any inappropriate operation of a gate in a circuit which results in an incorrect output is termed as cell faults (CF). Here the gates are referred to as a cell. The foundation of these faults is belongs to fault modeling in cellular logic arrays and therefore these faults can be simply called by cell faults. There are multiple anonymous ways of the occurrence of this kind of fault in the circuit, hence the calculation of its number is redundant.

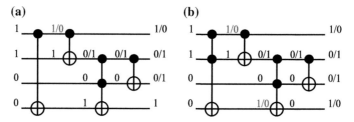

Fig. 4 Bit-flip faults (*Source* [48])

It can be noted that the occurrence of a type of fault in a circuit will results in flipping or inversion of bit values at the nodes after the gate where a fault occurrence has been taken place in the circuit. The alteration of these bits will affect single or multiple values at subsequent stages of the circuit and obviously cultivated toward the output. This kind of situation is referred to as a bit-flip fault can be termed as bit faults. When the value of a single bit is distorted, it can be a single-bit fault and if multiple values are flipped, it can be called as multiple bit faults. Figure 4a and b depict the patterns of changes of bit values because of respective faults in the circuit. Where, the propagation S-a-0 fault in red color for better understanding. The un-faulty/faulty bit values can be seen against each wire where the exactly these faults have affected.

It also can be concluded that the bit faults are meant for online testing as these faults detection will result in the detection of all types of fault models. The number of single-bit faults can be given by ($\sum_{i=1}^{G} g_i + n$). It diminishes the requirements of designing a separate hardware/method for the detection of a given type of fault. The design for test complexities can be reduced as a consequence.

3 Fault Identification

Every fault model has its own role to change the behavior of the circuit. Their identification is based on the type of gates used and the test vector which changes the input–output logic values. For instance, if an input vector is not able to interrupt the functioning of a gate in the circuit, it cannot be used for identification of any faults. An applied test vector to the inputs of the circuit alters one or more bit logic values of the input wires of the gates and subsequent levels contained by it. The identification procedures for the different type of faults in MCT and MCF gates are explained as follows:

3.1 Fault Identification for MCT Gates

Considering an n wire MCT gate having $(k_1, k_2, \ldots, k_m)$ control inputs and target input T. Note that, there cannot be any bridging and cross-point faults in a NOT gate. Respective deterministic methodologies for the identifications of faults are explained below.

SAF Identification

Single stuck-at faults in MCT circuit is given by $\{n + \sum_{i=1}^{GC}(k_{m_i}) + 2GC\}$, where, GC is number of gates and k_{m_i} is control inputs of ith gate of the circuit. Assuming logic 0 and 1 values at all the fault sites, the n dimensional test vector of size 2 given by $(0, 0, \ldots, 0)(1, 1, \ldots, 0)$ defines a test vector for the detection of all single and multiple type stuck-at faults of an MCT gate.

BrF Identification

Bridging faults is dependent on total number of wires in the circuit. All single intra-level bridging fault for an MCT gate is given by $\{2(GC + 1) \times^n C_2\}$. The detection is done by assuming complementary values between the two wires at every level of the circuit. The n dimensional test vector of size n, produced by shifting 0 value from first wire to next until it reaches to the end of test vector, given as $(0, 1, \ldots, 1), (1, 0, \ldots, 1), \ldots (1, 1, \ldots, 0)$ is complete for all single intra-level bridging faults of an MCT gate of a circuit.

MGF Identification

Single missing gate faults is equal to the gates present in the circuit. The detection principle for this type of fault is to assign logic values 1 to the control inputs and 1 (or 0) value to the target input of the gate in the circuit. The n-dimensional single test vector given as $\{k_1, k_2, \ldots k_m, T\} = \{(1, 1, \ldots 1, 1)\}$ (or $\{(1, 1, \ldots 1, 0)\}$) is complete for its detection.

CPF Identification

The number of single cross-point faults, either appearance or disappearance type, occurred in the circuit is given by $N(n - 1)$. The detection is achieved by assigning the combination of n test vectors keeping logic 1 to the $m - 1$ control inputs and 1 value to target input of each gate at distinct levels of the circuit. Assuming logic 0 to the control input where the fault has been occurred, making rest all control at logic 1 and 1 (or 0) value to target input of the gate. The n dimensional test vector of size $(n - 1)$ given as $\{k_1, k_2, \ldots k_m, T\} = \{(0, 1, \ldots 1, 1), (1, 0, \ldots 1, 1), \ldots (1, 1, \cdots 0, 1)$ is complete for the detection of all CPF of the gate for $n \geq 2$.

CF Identification

The 2^n greedy permutation fix the detection of this type of fault and all n dimensional test vectors of size 2^n are required to detect all the cell faults in the circuit.

3.2 Fault Identification for MCF Gates

Considering an n wire MCF gate having $(k_1, k_2, \ldots, k_m)$ control inputs and target inputs T_1 and T_2. Note that, there will be no cross-point faults in a swap gate and no single wire MCF gate is available. Also, the test principles used for traditional and MCT based logic circuits can be applied for the detection of stuck-at faults, single intra-level bridging faults, and cell faults. Respective deterministic methodologies are explained below.

MGF Identification

Single missing gate faults in MCF circuit is equal to the sum of gates available in the circuit. The detection principle for this type of fault is to assign logic values 1 to the control inputs and complementary values to the two target inputs of the gate in the circuit. The n dimensional single test vector given as $\{k_1, k_2, \ldots k_m, T_1, T_2\} = \{(1, 1, \ldots 1, 0, 1)\}$ is complete for its detection.

CPF Identification

The number of single cross-point faults, either appearance or disappearance type, occurred in the circuit is given by $N(n - 2)$. The detection is achieved by assigning the combination of n test vectors keeping logic 1 to the $m - 2$ control inputs and complementary values to target input of each gate at every level of the circuit. Assuming logic 0 to the control input where the fault has been occurred, making rest all control at logic 1 and complementary values to the target inputs of the gate. The n dimensional test vector of size $(n - 2)$ given as $\{k_1, k_2, \ldots k_m, T_1, T_2\} = \{(0, 1, \ldots 1, 0, 1), (1, 0, \ldots 1, 0, 1), \ldots (1, 1, \cdots 0, 0, 1)$ is complete for the detection of all CPF of the gate for $n \geq 3$.

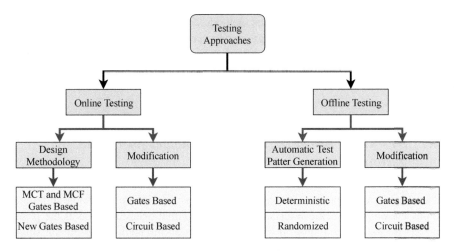

Fig. 5 Classification of testing approaches

4 Testing Approaches

Testing of the reversible circuit gaining grounds since more than a decade, where the researchers have been forwarded about the kind of faults that can be happed and developing their testing strategies in this emerging field. A collective information of testing approaches from the literature is provided in this section with respect to cost metrics and test complexity. The possible categorization of testing approaches for reversible circuits that have been proposed so far [49–51], are shown in Fig. 5. A brief about the contribution in these approaches are described in this section. It can be noted that mostly all the online testing approaches utilize parity checking technique. However, offline approaches exploit the bijective mapping property of reversible function. Number of inputs, gates, quantum cost, ancillas, and garbage are considered to evaluate the performance of respective testing approaches. Three more measures are included for the analysis of testing approaches: test size (s), execution time (t), and fault coverage (FC) [51]. However, test size and execution time are used in case of ATPG approaches.

4.1 MCT and MCF Gates-Based Design Methodologies for Online Testing

The MCT and MCF are the fundamental gates and the final quantum decomposition is based on them. The design complexity of testable and quantum circuits are proven lesser due to this reason [52]. The design methodologies which provide built-in testability features includes a twofold MCT gate placement, design with MCF gates and design with mixed MCTF gates methodologies were proposed [11, 53]. These methodologies depict a novel design paradigm that relies on the placement of MCT and MCF gates to generate a desired Boolean function. The placement methodology produce parity preserving circuits and are meant for single bit fault detection, which are called as soft errors in a broad sense.

4.2   New Gates Based Design Methodologies for Online Testing

The objective of proposing testable gates (new gates) is to incorporate some additional input and output to achieve testability. Using minimal operating costs is the prime targets for their formulation. It can be noted that all the methods are meant single bit faults detection inside a circuit. These new gates are 4×4 R1 and R2 whose combination formulates a testable block TB [4, 5] and testable gate OTG which can be used to produce testable cell CTSG. Two self-testable dual-rail coding scheme based gates are also presented [8] to remove the requirements of additional dual-rail checkers in prior methodologies. The gates LIG and FIG are introduced to establish testable information at the output for MGF faults without using parity checking [9]. The method for multi-bit errors is also formulated using concurrent error detection scheme [10].

4.3   Gate-Based Modification Methodologies for Online Testing

The gates of the circuit are modified to achieve testability in these type of approaches. A modification of a reversible gate procedure to make them testable form is first adopted by the use of ETG (Extended Toffoli Gate) [54] is adopted. However, ETGs are formulated over photonics, the methodology utilizes two additional CNOT gates per MCT gate. Another method exploits the technique of cascading several stages of an identity gate to renovate a gate into a testable reversible cell TRC. Both of the methodologies exploits the phenomenon of parity preservation and generation for single bit fault detection in reversible circuits. A method of gates conversion for testability is presented utilizing the property of parity preserving gates rather converting for the same [48]. A methodology for the modification of Toffoli and Peres gates in corresponding testable form is also presented which ensures the nearly all multiple-bit faults detection [16].

4.4 Circuit-Based Modification Methodologies for Online Testing

However, the circuit is modified when the modified gates are used for their formulation. In this particular approach complete circuit is circuit using the same testable viewpoints at all levels or some levels of the circuit. In this regard, a method of gates cascading was proposed for the modification of MCT circuits [55]. In this method, a gate of same size is cascaded after each gate with same control inputs and target on a new wire. This modification produces parity preserving reversible circuits where the detection of single bit faults can be achieved using two arrays of CNOT gates. The method is also proven better in performance when implemented using quantum cellular automata (QCA) platform.

4.5 Deterministic ATPG Approaches for Offline Testing

Utilizing the concepts described in Sect. 3 for various types of faults detection, several algorithms are proposed. Complete test sets for detection of stuck-at and cell faults are formulated using practical heuristic over an integer level program (ILP) using binary variables [17, 18]. The test sets by obtained in lesser execution time and proven for minimal test size. An exact ATPG is developed over the emerging ion-trap quantum computing technology for obtaining minimal test set for missing gate faults [19]. Principle of comparison of change in output due to missing gates based and the concepts of Boolean difference method, algorithms are also developed for single and multiple missing gate faults detection respectively [20, 21]. The test generation algorithm is formulated for the detection of bridging faults on the basis of proposed block division method [22]. An algorithm which produces a constant universal test vector (of size n) is also presented for single and multiple input bridging faults [23]. It is noted that, nearly all the presented methodologies are meant for MCT circuit.

4.6 Randomized ATPG Approaches

Creation of and modification of existing ATPGs by including random variables is another method, where researchers also proposed several solutions for achieving testability in MCT circuits. Solving an NP-hard problem for obtaining minimal size test set stuck-at fault detection in NCT circuits is proposed [28]. Greedy heuristic and exact branch and bound algorithm are utilized to detect the missing gate faults [29]. Ping-pong method is proposed to generate a test set for missing gate type of faults [30]. The detection of cross-point faults is also generalized by proposing a randomized ATPG [31].

4.7 Gates Based Modification Methodologies for Offline Testing

As per the authors knowledge and review of this area, transformation of a reversible to corresponding AND-EXOR irreversible circuit is the only method presented in the literature for the detection of single intra-level bridging faults [36]. MCT circuits are first decomposed in the irreversible PPRM AND-EXOR network. Faults are assumed between the wires of AND-EXOR circuit and detection strategy is applied at its different levels.

4.8 Circuit Based Modification Methodologies for Offline Testing

Following the identification approaches of the faults in reversible circuits, the circuit is modified in such a method that the applied test vector propagates till the last level of the circuit in this type of methodology. Two circuit modification methodologies are formulated for single and multiple stuck-at faults detection in MCT circuits. A universal test set (UTS) and a complete test set (CTS) test sizes of 2 are proposed along with the methodologies. An efficient adding a gate methodology is also presented for the detection and location of these faults with a minimal test set for their detection [56]. Techniques are realized by the inclusion CNOT gates for missing gate fault detection by single test vector [17, 29] and universal test set of size $(n + 1)$ [34, 35].

5 Summary

Diverse varieties of fault families in reversible logic circuits and an exclusive study of testable design advances for these faults are portrayed in this chapter. As a new technological change, an outstanding awareness is depicted by the researchers of the area for finding a solution for the notation and detection of these faults. Plentiful approaches were projected under in two extensive classifications to meet the challenge. The methodologies are alleged to coat almost all the faults and their sub kind by exploiting the properties of reversible gates and circuits. The objective is to minimize testing overhead which can be achieved by reducing the cost metrics utilized for testability. Following are the key points discussed in this chapter which includes objectives, notations, and result analysis for design and testing of reversible logic circuits:

References

1. Nielsen MA, Chuang IL (2011) Quantum computation and quantum information: 10th anniversary edition, 10th edn. Cambridge University Press, New York
2. Hurst SL (1998) VLSI Testing: digital and mixed analogue/digital techniques. The Institution of Electrical Engineers, London
3. Jha NK, Gupta S (2003) Testing of digital systems. Cambridge University Press, Cambridge
4. Vasudevan DP, Lala PK, Parkerson JP (2004) Online testable reversible logic circuit design using nand blocks. In: 19th IEEE international symposium on defect and fault tolerance in VLSI systems, 2004. DFT 2004. Proceedings, pp 324–331. https://doi.org/10.1109/DFTVS.2004.1347856
5. Vasudevan DP, Lala PK, Di J, Parkerson JP (2006) Reversible-logic design with online testability. IEEE Trans Instrum Meas 55(2):406–414. https://doi.org/10.1109/TIM.2006.870319
6. Hasan M, Islam AKMT, Chowdhury AR (2009) Design and analysis of online testability of reversible sequential circuits. In: 2009 12th international conference on computers and information technology, pp 180–185. https://doi.org/10.1109/ICCIT.2009.5407143
7. Thapliyal H, Vinod AP (2007) Designing efficient online testable reversible adders with new reversible gate. In: 2007 IEEE international symposium on circuits and systems, pp 1085–1088. https://doi.org/10.1109/ISCAS.2007.378198
8. Farazmand N, Zamani M, Tahoori MB (2010) Online fault testing of reversible logic using dual rail coding. In: 2010 IEEE 16th international on-line testing symposium, pp 204–205. https://doi.org/10.1109/IOLTS.2010.5560205
9. Zamani M, Tahoori MB (2011) Online missing/repeated gate faults detection in reversible circuits. In: 2011 IEEE international symposium on defect and fault tolerance in VLSI and nanotechnology systems, pp 435–442. https://doi.org/10.1109/DFT.2011.56
10. Thapliyal H, Ranganathan N (2010) Reversible logic based concurrent error detection methodology for emerging nanocircuits. In: 10th IEEE international conference on nanotechnology, pp 217–222. https://doi.org/10.1109/NANO.2010.5697743
11. Gaur HM, Singh AK (2016) Design of reversible circuits with high testability. Electron Lett 52(13):1102–1104. https://doi.org/10.1049/el.2016.0161
12. Nayeem NM, Rice JE (2011c) A simple approach for designing online testable reversible circuits. In: Proceedings of 2011 IEEE Pacific rim conference on communications, computers and signal processing, pp 85–90. https://doi.org/10.1109/PACRIM.2011.6032872
13. Nayeem NM, Rice JE (2011a) Online fault detection in reversible logic. In: 2011 IEEE international symposium on defect and fault tolerance in VLSI and nanotechnology systems, pp 426–434. https://doi.org/10.1109/DFT.2011.55
14. Nayeem NM, Rice JE (2013) Online testable approaches in reversible logic. J Electron Test 29(6):763–778. https://doi.org/10.1007/s10836-013-5399-3
15. Nayeem NM, Rice JE (2012) A new approach to online testing of tgfsop-based ternary toffoli circuits. In: 2012 IEEE 42nd international symposium on multiple-valued logic, pp 315–321. https://doi.org/10.1109/ISMVL.2012.57
16. Sen B, Das J, Sikdar BK (2012) A DFT methodology targeting online testing of reversible circuit. In: 2012 international conference on devices, circuits and systems (ICDCS), pp 689–693. https://doi.org/10.1109/ICDCSyst.2012.6188661
17. Patel KN, Hayes JP, Markov IL (2003) Fault testing for reversible circuits. In: Proceedings of the 21st VLSI test symposium, 2003, pp 410–416. https://doi.org/10.1109/VTEST.2003.1197682
18. Patel KN, Hayes JP, Markov IL (2004) Fault testing for reversible circuits. IEEE Trans Comput-Aided Des Integr Circuits Syst 23(8):1220–1230. https://doi.org/10.1109/TCAD.2004.831576
19. Polian I, Fiehn T, Becker B, Hayes JP (2005) A family of logical fault models for reversible circuits. In: 14th Asian test symposium (ATS'05), pp 422–427. https://doi.org/10.1109/ATS.2005.9

20. Kole DK, Rahaman H, Das DK, Bhattacharya BB (2011) Derivation of automatic test set for detection of missing gate faults in reversible circuits. In: 2011 international symposium on electronic system design (ISED). IEEE, pp 200–205

21. Mondal B, Kole DK, Das DK, Rahaman H (2014) Generator for test set construction of SMGF in reversible circuit by boolean difference method. In: 2014 IEEE 23rd Asian test symposium, pp 68–73. https://doi.org/10.1109/ATS.2014.24

22. Rahaman H, Kole DKKDK, Das DK, Bhattacharya BB (2007) Optimum test set for bridging fault detection in reversible circuits. In: 16th Asian test symposium (ATS 2007), pp 125–128. https://doi.org/10.1109/ATS.2007.91

23. Sarkar P, Chakrabarti S (2008) Universal test set for bridging fault detection in reversible circuit. In: 2008 3rd international design and test workshop, pp 51–56. https://doi.org/10.1109/IDT.2008.4802464

24. Babu HMH, Mia MS, Biswas AK (2017) Efficient techniques for fault detection and correction of reversible circuits. J Electron Test, pp 1–15. https://doi.org/10.1007/s10836-017-5679-4

25. Fujita M (2015) Detection of test patterns with unreachable states through efficient inductive-invariant identification. In: 2015 IEEE 24th Asian test symposium (ATS), pp 31–36. https://doi.org/10.1109/ATS.2015.13

26. Fujita M (2015) Automatic identification of assertions and invariants with small numbers of test vectors. In: 2015 33rd IEEE international conference on computer design (ICCD), pp 463–466. https://doi.org/10.1109/ICCD.2015.7357149

27. Burchard J, Erb D, Singh AD, Reddy SM, Becker B (2017) Fast and waveform-accurate hazard-aware sat-based TSOF ATPG. In: Design, automation test in Europe conference exhibition (DATE), 2017, pp 422–427. https://doi.org/10.23919/DATE.2017.7927027

28. Tabei K, Yamada T (2009) On generating test sets for reversible circuits. In: 2009 international conference on computer engineering systems, pp 94–99. https://doi.org/10.1109/ICCES.2009.5383305

29. Hayes JP, Polian I, Becker B (2004) Testing for missing-gate faults in reversible circuits. In: 13th Asian test symposium, pp 100–105. https://doi.org/10.1109/ATS.2004.84

30. Zamani M, Tahoori MB, Chakrabarty K (2012) Ping-pong test: Compact test vector generation for reversible circuits. In: 2012 IEEE 30th VLSI test symposium (VTS), pp 164–169. https://doi.org/10.1109/VTS.2012.6231097

31. Zhong J, Muzio JC (2006) Analyzing fault models for reversible logic circuits. In: 2006 IEEE international conference on evolutionary computation, pp 2422–2427. https://doi.org/10.1109/CEC.2006.1688609

32. Chakraborty A (2005) Synthesis of reversible circuits for testing with universal test set and c-testability of reversible iterative logic arrays. In: 18th international conference on VLSI design held jointly with 4th international conference on embedded systems design, pp 249–254. https://doi.org/10.1109/ICVD.2005.158

33. Ibrahim M, Chowdhury AR, Babu HMH (2008) Minimization of CTS of k-CNOT circuits for SSF and MSF model. In: 2008 IEEE international symposium on defect and fault tolerance of VLSI systems, pp 290–298. https://doi.org/10.1109/DFT.2008.38

34. Rahaman H, Kole DK, Das DK, Bhattacharya BB (2008) On the detection of missing-gate faults in reversible circuits by a universal test set. In: 21st international conference on VLSI design (VLSID 2008), pp 163–168. https://doi.org/10.1109/VLSI.2008.106

35. Rahaman H, Kole DK, Das DK, Bhattacharya BB (2011) Fault diagnosis in reversible circuits under missing-gate fault model. Comput Electr Eng 37(4):475–485. https://doi.org/10.1016/j.compeleceng.2011.05.005

36. Chakraborty A (2010) Testing of bridging faults in AND-EXOR based reversible logic circuits. CoRR arXiv:abs/1009.5098

37. Bubna M, Goyal N, Sengupta I (2007) A DFT methodology for detecting bridging faults in reversible logic circuits. In: TENCON 2007 - 2007 IEEE region 10 conference, pp 1–4. https://doi.org/10.1109/TENCON.2007.4428915

38. Parhami B (2006) Fault-tolerant reversible circuits. In: 2006 fortieth asilomar conference on signals, systems and computers, pp 1726–1729. https://doi.org/10.1109/ACSSC.2006.355056

39. Avizienis A (1978) Fault-tolerance: the survival attribute of digital systems. Proc IEEE 66(10):1109–1125. https://doi.org/10.1109/PROC.1978.11107
40. Mathew J, Singh J, Taleb AA, Pradhan DK (2008) Fault tolerant reversible finite field arithmetic circuits. In: 2008 14th IEEE international on-line testing symposium, pp 188–189. https://doi.org/10.1109/IOLTS.2008.35
41. Vasudevan DP, Lala PK, Parkerson JP (2005a) The construction of a fault tolerant reversible gate for quantum computation. In: 5th IEEE conference on nanotechnology, 2005, vol 1, pp 112–115. https://doi.org/10.1109/NANO.2005.1500705
42. Vasudevan DP, Lala PK, Parkerson JP (2005) Fault tolerant quantum computation with new reversible gate. In: Technical proceedings of the 2005 NSTI nanotechnology conference and trade show, vol 3, pp 744–747
43. Boykin PO, Roychowdhury VP (2005) Reversible fault-tolerant logic. In: 2005 international conference on dependable systems and networks (DSN'05), pp 444–453. https://doi.org/10.1109/DSN.2005.83
44. Zamani M, Farazmand N, Tahoori MB (2011) Fault masking and diagnosis in reversible circuits. In: 2011 sixteenth IEEE European test symposium, pp 69–74. https://doi.org/10.1109/ETS.2011.19
45. Abdessaied N, Amy M, Soeken M, Drechsler R (2016) Technology mapping of reversible circuits to clifford+t quantum circuits. In: 2016 IEEE 46th international symposium on multiple-valued logic (ISMVL), pp 150–155. https://doi.org/10.1109/ISMVL.2016.33
46. Amy M, Maslov D, Mosca M, Roetteler M (2013) A meet-in-the-middle algorithm for fast synthesis of depth-optimal quantum circuits. IEEE Trans Comput-Aided Des Integr Circuits Syst 32(6):818–830. https://doi.org/10.1109/TCAD.2013.2244643
47. iNEMI (2015) Roadmap executive summary highlights. International electronics manufacturing initiative
48. Gaur HM, Singh AK, Ghanekar U (2017) Testable design of reversible circuits using parity preserving gates. In: IEEE design and test
49. Gaur HM, Singh AK, Ghanekar U (2015) A review on online testability for reversible logic. Procedia Comput Sci 70:384–391
50. Gaur HM, Singh AK, Ghanekar U (2016a) A comprehensive and comparitive study on online testability in reversible logic. Pertanika J Sci Technol 24(2):245–271
51. Gaur HM, Singh AK, Ghanekar U (2018) Offline testing of reversible logic circuits: an analysis. Integration 62:50–67. https://doi.org/10.1016/j.vlsi.2018.01.004, http://www.sciencedirect.com/science/article/pii/S0167926016301341
52. Gaur HM, Singh AK, Ghanekar U (2018b) In-depth comparative analysis of reversible gates for designing logic circuits. Procedia Comput Sci 125:810–817
53. Gaur HM, Singh AK, Ghanekar U (2018d) Reversible circuits with testability using quantum controlled not and swap gates. Indian J Pure Appl Phys 56(7):529–532
54. Mahammad SN, Veezhinathan K (2010) Constructing online testable circuits using reversible logic. IEEE Trans Instrum Meas 59(1):101–109. https://doi.org/10.1109/TIM.2009.2022103
55. Gaur HM, Singh AK, Ghanekar U (2016b) A new dft methodology for k-cnot reversible circuits and its implementation using quantum-dot cellular automata. Optik - International Journal for Light and Electron Optics 127(22):10593–10601. https://doi.org/10.1016/j.ijleo.2016.08.072, http://www.sciencedirect.com/science/article/pii/S003040261630941X
56. Gaur HM, Singh AK, Ghanekar U (2018a) Design for stuck-at faults testability in mct based reversible circuits. Def Sci J 68(4):381–387

Detection and Identification of Gate Faults in Reversible Circuit

B. Mondal, C. Bandyopadhyay, A. Bhattacharjee and H. Rahaman

Abstract In recent time, efficient implementation of reversible logic circuits has come out as an important research area before the design industry. With the advancement in reversible logic synthesis, developing mechanism for identification of faults finds importance. Though there exist well-known testing techniques, but developing improved testing algorithms is the need of the hour. Aiming to develop efficient testing technique, here in this work, we show an improved testing scheme based on Boolean logic function. Two different testing approaches are presented here, where in the first work, by using Boolean difference method SMGFs in reversible circuit are tracked successfully, where a test vector generator is derived to find the faults. In the second work, a Reed–Muller (RM) form based testing approach is developed that not only detects the faults but also locates the exact position of the faulty area. A limitation for the second testing scheme is that it can only be employed over a specific type of reversible circuit known as Exclusive-Or Sum-Of-Product (ESOP) design. Both the testing techniques have been executed over different benchmark suites and a comparative study with state-of-the-art testing approaches have been included in the work.

B. Mondal (✉) · C. Bandyopadhyay · A. Bhattacharjee · H. Rahaman
Department of Information Technology, Indian Institute of Engineering Science and Technology, Shibpur 711103, India
e-mail: bappa.arya@gmail.com

C. Bandyopadhyay
e-mail: chandanb.iiest@gmail.com

A. Bhattacharjee
e-mail: anirbanbhattacharjee330@gmail.com

H. Rahaman
e-mail: hafizur@it.iiests.ac.in

© Springer Nature Singapore Pte Ltd. 2020
A. K. Singh et al. (eds.), *Design and Testing of Reversible Logic*, Lecture Notes in Electrical Engineering 577, https://doi.org/10.1007/978-981-13-8821-7_10

169

1 Introduction

Heat dissipation is considered as the essential concern in modern day's VLSI circuit. As per Launder's principles [1], loss of information generates $KTlog_2 joule$ amount of heat, where k is Boltzmann constant and T is absolute temperature. Hence, alternative technology is required so that heat generation can be minimized in the circuit. Bennet [2] claimed that dissipation of energy can be made zero only when the circuit is constructed with reversible gates. Therefore, reversible logic design is considered as the prerequisite needed to minimize heat dissipation during logic computation. On the other side, as the quantum circuit [3] follows the principle of reversibility, the implementation of quantum functionality using reversible circuit is possible. Reversible circuit not only has the dominance in the field of quantum circuit design, but it too has applications in adiabatic computing [4, 5], Cryptography and Optical Computing. In the last couple of years, several progresses have been made on efficient design strategies of reversible circuit.

Synthesis algorithms have been developed for making the designs of reversible circuit generic. But, not only designing the cost-efficient circuits get the high importance but simultaneously developing testing algorithms [6–11] for checking the correctness of such designs find popularity. In recent time, some promising works on efficient testing strategies have been developed where improved algorithms are formulated to make the testing process easier.

Here, in this work, we show two different approaches to find faults in reversible circuit. In the first work, a Boolean difference-based testing technique is developed, where a Boolean generator is formulated to produce test vectors and then the faults are tracked. This approach is very generic as it can be employed over any type of circuits. The second testing scheme is not very generic like the first one as it can only operate over ESOP-based designs. In this testing scheme, the functional power of Reed–Muller expression is used to find and locate the faults.

The remaining portion of the article is formulated as follows: preliminaries associated with reversible testing are stated in Sect. 2. Section 3 summarizes previous research works on reversible testing. The developed methodologies are presented in Sect. 4. The experimental data of our work are summarized in Sect. 5. Finally, the chapter is concluded in Sect. 6.

2 Preliminaries

2.1 Reversible Circuits and Gates

Definition 1 A circuit C_{nf} over a set of circuit lines $L = \{c_1, c_2, ..., c_n\}$ is said to be reversible if it satisfies the following three criteria:

(i). input (m) lines are equal with the output (n) lines

Fig. 1 a 2-control Toffoli gate, **b** *CNOT* gate, **c** *NOT* gate

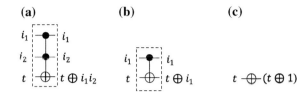

(ii).   if the circuit is fan-out free

(iii).   circuit consists of reversible gates only.

Definition 2 A reversible gate G can be described as G(C; T), where parameters C, T represents the control and target connection inputs. In that gate G, the control input set C may contain an empty value but the set T must have a minimum of one target line in such that $C \cap T = \Phi$.

In classical circuit different logic gates are used to implement a circuit, similarly there are well-known reversible gates like Toffoli [12], Fredkin [13], Feynman [14] that are used to construct reversible circuits. Some examples of reversible gates are depicted in Fig. 1.

2.2   ESOP-Based Design

A reversible circuit may have different designs and such variations in design depend on the type of algorithm deployed or heuristic employed. Among the several design models, due to the scalable feature property, ESOP (Exclusive Sum-Of-Products) [15]-based representation has been found as one of the widely used design model for reversible circuit. Now in the following, we introduce this special type of circuit.

ESOP can be represented in the form of Sum-Of-Products (SOP) form except that the SOP product terms which are separated by '+' operator, it is separated by "$\oplus$" operator. To express any n-input, m-output reversible function in ESOP representation, it requires $(n + m)$ numbers of input lines in the circuit, where n represents the control set and the rest m lines operate as functional output lines.

Cube list a special data structure from which the ESOP designs are formed. Such cube list contains the detail gate specification and control structure for the ESOP circuit. Each cube in the cube list denotes a gate in the design. For an ease of understanding, cube list and its corresponding ESOP expression are given in Fig. 2a and b, where it can be seen that each of the cubes from the list has been mapped to an equivalent gate and finally a complete design is formed Fig. 2b.

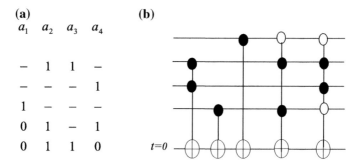

Fig. 2 **a** Cubelist for $f(a_1, a_2, a_3, a_4) = a_2a_3 \oplus a_4 \oplus a_1 \oplus \bar{a}_1a_2a_4 \oplus \bar{a}_1a_2a_3\bar{a}_4$, **b** ESOP expression of Fig. 2a

2.3 Reed–Muller Form

For efficient design and testing of reversible circuit, the concept of Boolean algebra operator is used widely. The modulo-2 arithmetic is applied and any Boolean function can be realized using this algebra. For implementation purpose, Reed–Muller expansion can be represented using sum-of-products expression of modulo-2 arithmetic. Reversible circuit based on module-2 expansion can be realized using only exclusive OR (EXOR) gates. For any Boolean function $f(x_1x_2 \ldots x_n)$, it is expressed in the form of Reed–Muller expansion [16, 17].

PPRM: The positive polarity based Reed–Muller (PPRM) expression can be realized in the form of an EXOR canonical sum-of-products expression in which each variable represents a positive polarity (un-complimented). A PPRM expression of n variable can be expressed as

$f(x_1x_2 \ldots x_n) = a_0 \oplus a_1x_1 \oplus a_2x_2 \oplus a_3x_1x_2 \oplus \ldots \oplus a_{2^n-1}x_1x_2 \ldots x_n$, where $a_i \in \{0, 1\}$. The variable a_i in the above expression represents coefficient vector, where x_i denotes input variables. If the coefficient becomes zero, then the corresponding product term is not present in the PPRM expression otherwise the product term is included in the given expression.

FPRM: In fixed polarity Reed–Muller (FPRM) expression, the n variable function can be represented as

$f(x_1x_2 \ldots x_n) = a_0 \oplus a_1\dot{x}_1 \oplus a_2\dot{x}_2 \oplus a_3\dot{x}_1\dot{x}_2 \oplus \ldots \oplus a_{2^n-1}\dot{x}_1\dot{x}_2 \ldots \dot{x}_n$ where $a_i \in \{0, 1\}$ and $\dot{x} \in \{x, \bar{x}\}$, where x denoted un-complemented literals and $\bar{x}$ denotes complemented literals.

GRM/MPRM: The Mixed polarity Reed–Muller (MPRM) expression can be considered as the generalization of FPRM expression in which there is hardly any limitation in the polarity of each variable. The Boolean function having n variable can be represented by a number of $2^{n2^{n-1}}$ GRM form. The MPRM expression for n variable function is represented as

$f(x_1 x_2 \ldots x_n) = a_0 \oplus a_1 \dot{x}_1 \oplus a_2 \dot{x}_2 \oplus a_3 \dot{x}_1 \dot{x}_2 \oplus \ldots \oplus a_{2^n-1} \dot{x}_1 \dot{x}_2 \ldots \dot{x}_n$ where $a_i \in \{0, 1\}$ and $\dot{x} \in \{x, \bar{x}\}$.

The association among various Reed–Muller configurations can be represented as PPRM $\subset$ FPRM $\subset$ GRM/MPRM.

2.4 Different Fault Models and Their Properties

Faults in a circuit may originate due to several reasons [18–20]. Depending on the type of errors, there are four types of faults such as—single missing gate fault (SMGF), repeated gate fault (RGF), partial missing gate fault (PMGF) and multiple missing gate fault (MMGF).

Definition 3 Complete disappearance of a gate from a given circuit results in a single missing gate fault.

The Fig. 3a shows an SMGF in a benchmark circuit *ham3\design#*, where the faulty area has been marked with a dotted box. In the dotted region, the first 2-CNOT gates are missing. A SMGF fault can be detected by providing value 1 to all the control line set of the gate and either 0 or 1 at the target node of corresponding gate.

Definition 4 A fault can be considered as repeated gate fault if the same gate consecutively reappears in the design and may change the functionality of the circuit.

In Fig. 3b, the RGF is shown in the first gate of the *ham3\design#1*. It can be ascertained that if a gate reappears even number of times then its effect becomes equivalent to an SMGF fault while odd occurrence makes the RGF fault as redundant indicating the circuit functionality remains unchanged.

Definition 5 Disappearance of control connection input of a gate results in a fault in a circuit known as partial missing gate fault (PMGF).

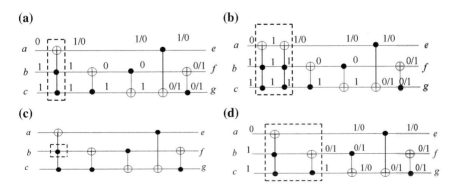

Fig. 3 **a** SMGF fault in circuit*ham3\design#*, **b** RGF in *ham3\design#1* circuit, **c** PMGF in circuit*ham3\design#1*, **d** MMGF in *ham3#design#1*

In Fig. 3c, a PMGF fault is shown in the first gate of the circuit. It is considered that a PMGF fault can only be uncovered by applying value 0 to at the missing control inputs and a value 1 is set for other control lines.

Definition 6 Missing of two or more successive gates from the circuit generates a fault known as multiple missing gate fault (MMGF).

Consider the circuit shown in Fig. 3d, where the first two gates enclosed within the dotted box are missing.

Still so far we have seen the fundamentals related to reversible circuit and its associated fault models. Now, here we are discussing some of the works in testing and highlighting their contributions.

3   Related Works and Contributions

An efficient approach of stuck-at-fault detection by the adaptive tree-based technique is proposed in [6]. Here, in this technique at first the fault in half of the circuit is detected and then by applying the reversible property, a mirror image of the remaining part of the circuit is developed to detect faults in the remaining part of the circuit.

Furthermore, a generalized approach of "stuck-at" fault detection for all k-CNOT based circuit has been reported in [7], where a universal test of size 3 has been used for the fault detection.

Taking one step more, a novel technique for fault detection is shown in [8], where an (n × n) reversible circuit constructed with k-CNOT gates is tested for possible faults. In this method, a testable design has been developed by copying gates along with an additional line. Though some overhead is incurred to transform the original circuit into the testable form, but the modified testable design becomes very sufficient and easy to detect all existing fault models in the given circuit.

To make the faults detection easier, not only fault specific test vectors have been generated but the way of making simpler testable designs also have been explored and such a work is reported in [9], where extra inputs and additional k-CNOT gates are added in the design to make the design testing friendly. This modified testing design methodology determines a universal test set of size $(n + 2)$ and thereby identifies the said faults in the given circuit.

Instead of applying huge number of test vectors in fault detection and also to reduce the design complexity, testing of SMGFs and RGFs and MMGFs in reversible circuit with minimum number of test vectors is presented in [10].

4 Proposed Testing Methods

Here we state both the testing schemes with examples. The first testing scheme is based on Boolean difference method, whereas the second one relies on Reed–Muller expansion.

4.1 Boolean-Based Testing Method1

Here we state the technique to determine the existence of SMGF in a circuit. The proposed approach is divided into three phases. At first, Boolean difference of the circuit is computed for each of the missing gate. In the second phase, a Boolean generator for the entire circuit is constructed and in the third phase, test vectors are constructed from the Boolean generator which finally checks the presence of SMGF in the circuit. All three phases are stated next in detail.

4.1.1 Computation of the Boolean Difference for Missing Gate Faults

Let us assume a reversible circuit having n number of input lines and N number of gates.

Definition 7 The Boolean expression generated at the jth line of a fault-free circuit is known as the OriginalExpression$_j$, where $(0 \leq j \leq n - 1)$. For any given reversible circuit with n inputs can be represented by OriginalExpression$_0$, OriginalExpression$_1$, OriginalExpression$_{n-1}$.

Definition 8 For a faulty circuit, the Boolean expression produced at the jth line as a result of the gate missing from the ith level of circuit can be represented as the FaultyExpression$_{i,j}$, where $(0 \leq i \leq N - 1)$, $(0 \leq j \leq n - 1)$.

For any reversible circuit with n lines can be expressed as *FaultyExpression$_{i,0}$*, *FaultyExpression$_{i,1}$*, …, *FaultyExpression$_{i,n-1}$* for the detected fault occurring at ith gate. The computed Boolean expression of the jth line in the faulty circuit may not be identical to the one generated at the jth line of the corresponding fault-free circuit.

Boolean difference method [21] is basically used to determine the complete test set for detecting stuck-at faults. We have employed the same it in reversible circuit for identifying SMGF faults.

Definition 9 The Boolean difference $\left(\frac{dF_j}{dG_i}\right)$ estimated at the jth line for the gate missing from the ith level can be expressed as $\frac{dF_j}{dG_i} = F_j^o \oplus F_j^i$, where F_j^o is the OriginalExpression$_j$, F_j^i is the FaultyExpression$_{i,j}$, $(0 \leq i \leq N - 1)$ and $(0 \leq j \leq n - 1)$.

Definition 10 The Boolean difference resulted due to removal of gate located at ith level is represented as $\frac{dF}{dG_i}$ which can be determined as follows:

$$\frac{dF}{dG_i} = \sum \frac{dF_j}{dG_i} = \left(\frac{dF_0}{dG_i}\right) + \left(\frac{dF_1}{dG_i}\right) + \cdots + \left(\frac{dF_{n-1}}{dG_i}\right), (0 \le i \le N-1), (0 \le j \le n-$$

In this way, at first the Boolean difference for the missing of each gate in the circuit is computed and then the Boolean difference of the circuit is constructed using the Boolean difference for each gate of the given circuit.

Definition 11 The Boolean generator can be defined as the Boolean expression needed for the test set construction so as to identify all the possible SMGFs in the circuit.

Lemma 1 *For a reversible circuit of n number input lines and N number of gates, then computation of individual Boolean difference* $\left(\frac{dF}{dG_i}\right)$ *enables to identify SMGF at the ith level.*

Proof For the reversible circuit with n lines and N gates, the Boolean difference $\left(\frac{dF}{dG_i}\right)$ for detecting the gate missing at the ith level is computed as:

$$\frac{dF}{dG_i} = \sum \frac{dF_j}{dG_i}, (0 \le i \le N-1), (0 \le j \le n-1) \tag{1}$$

Σ denotes the OR operation and

$$\frac{dF_j}{dG_i} = F_j^o \oplus F_j^i, (0 \le i \le N-1), (0 \le j \le n-1) \tag{2}$$

From the expression given at Eq. (1), it can be noticed that the possible values for $\frac{dF}{dG_i}$ can be either 0 or some other Boolean representation. Furthermore, it can be determined that the estimated result of $\frac{dF}{dG_i}$ becomes zero provided each of the terms $\frac{dF_j}{dG_i}$ evaluates to "0" as logical OR is being performed between any two successive terms of $\frac{dF}{dG_i}$.

By analyzing the expression given in Eq. (2), it can be observed that the term $\frac{dF_j}{dG_i}$ returns the value "0", if both F_j^o and F_j^i are becomes identical only if the circuit is fault free. This in turns suggests that, if the expression $\frac{dF}{dG_i}$ returns 0 then the circuit is said to be fault free.

For any SMGF in the circuit, it is obvious that the output expression obtained at any of the n input lines varies with the one derived for fault-free circuit. It means that at least a single line say j must be present for which the terms F_j^o and F_j^i turns out to be different due to existence of fault. It implies that $\frac{dF}{dG_i}$ cannot be equal to 0 in the faulty reversible circuit as logical OR is implemented between the consecutive terms $\frac{dF}{dG_i}$ and $\frac{dF_j}{dG_i}$.

4.1.2 Implementation of Boolean Generator for Detection of Single Missing Gate Fault

In the proposed method, the Boolean generator of the given circuit is estimated using the function $B_{Gen} = \frac{dF}{dG_0} \wedge \frac{dF}{dG_1} \wedge \cdots \wedge \frac{dF}{dG_{N-1}}$, where $\wedge$ represents the logical AND operation and N represents number of gates in the circuit. For BGen $\neq 0$, the BGen is minimized to construct the Boolean generator of the circuit. For BGen $= 0$, the expressions B_{Gen}^1 and B_{Gen}^2 need to be computed to determine the Boolean generator of the given circuit. The computation method of the expressions B_{Gen}^1 and B_{Gen}^2 are as follows:

Let S = (so, s1, ... , sN − 1) be the set of all $\frac{dF}{dG_i}$, where si $= \frac{dF}{dG_i}$ for ($0 \leq$ i $\leq$ N − 1). Let S1 $\subseteq$ S contains maximum number of si's such that $B_{Gen}^1 =$ Si1∧Si2∧ $\cdots$ ∧Sik $\neq 0$, where Sik $\in$ S1. B_{Gen}^2 are remaining si's of S. For each of the B_{Gen}^2, the $B_{Gen}^1 = 0$.

After the formation B_{Gen}^1 and B_{Gen}^2, the expression B_{Gen}^1 is upgraded to compute the final form of the Boolean generator as follows:

(i) If any term of B_{Gen}^2 completely matches or is subset of any term of B_{Gen}^1, no need to upgrade the B_{Gen}^1, else compare different terms of the B_{Gen}^1 with each term of the B_{Gen}^2 and upgrade the B_{Gen}^1 as: $B_{Gen}^1 = B_{Gen}^1 +$ highest matching term [i], i $= 0$ to (number of highest matching term − 1). Repeat the same procedure between upgraded B_{Gen}^1 and other available B_{Gen}^2, if exist.

(ii) The minimized form of B_{Gen}^1 is the Boolean generator of the circuit.

4.1.3 Test Vector Construction from the Boolean Generator of the Circuit

The minterms and their corresponding decimal values for the generator are calculated and the collection of those decimal values is the test set of the circuit that will be used for the detection of SMGF.

Example 1 The *ham3tc* benchmark circuit of Fig. 4a is considered here. The circuits consisting of three lines ($n = 3$) and five gates ($N = 5$).

As per the first phase of the proposed technique, output expressions *Original-Expression$_0$*, *OriginalExpression$_1$* and *OriginalExpression$_2$* are computed from the fault-free circuit of Fig. 4a.

The output expression generated for each circuit line is $OriginalExpression_0 = (a \oplus b \cdot c)$, $OriginalExpression_1 = ((b \oplus c) \oplus ((c \oplus (b \oplus c)) \oplus (a \oplus b \cdot c)))$, OriginalExpression$_2 = ((c \oplus (b \oplus c)) \oplus (a \oplus b \cdot c))$.

Now, let us assume that the gate G_0 at the *level$_0$* is removed from the circuit. Hence, the circuit *ham3tc* becomes a faulty circuit (as shown in Fig. 4b) and its output expressions are FaultyExpression0,0, FaultyExpression0,1, FaultyExpression0,2.

(a)

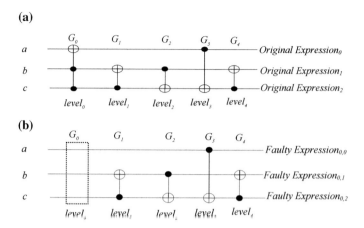

Fig. 4 **a** Fault free *ham3tc* circuit, **b** testable circuit for *SMGF* (Faulty *ham3tc* circuit where dotted box indicates missing of that gate)

Each expression of the faulty circuit are FaultyExpression$_{0,0}$ = (a), *FaultyExpression$_{0,1}$* = ((b ⊕ c) ⊕ ((c ⊕ (b ⊕ c) ⊕ (a))), FaultyExpression0,$_2$ = ((c ⊕ (b ⊕ c)) ⊕ (a)).

After finding out fault free expression of each line and the faulty expression of each line of the circuit, the Eqs. 1 and 2 as discussed earlier are used to compute the Boolean difference of the circuit. The Boolean difference of the circuit for the missing of the gate G0 is as $\frac{dF}{dG_0}$ = $\left(\frac{dF_0}{dG_0}\right)$ + $\left(\frac{dF_1}{dG_0}\right)$ + $\left(\frac{dF_2}{dG_0}\right)$, $\frac{dF}{dG_0}$ = (OriginalExpression0 ⊕ FaultyExpression0,0) + (OriginalExpression1 ⊕ FaultyExpression$_{0,1}$) + (OriginalExpression2 ⊕ FaultyExpression0,2) = ((a⊕b·c)⊕a)+(((b⊕c)⊕((c⊕(b⊕c)) ⊕(a ⊕ b · c))) ⊕ ((b ⊕ c) ⊕ ((c ⊕ (b ⊕ c) ⊕ (a))))) +(((c ⊕ (b ⊕ c)) ⊕ (a ⊕ b · c)) ⊕((c ⊕ (b ⊕ c)) ⊕ (a))) = bc

Similarly, each gate is removed at a time and the Boolean difference of the given circuit for the remaining gates is computed.

So, $\frac{dF}{dG_1}$ = c, $\frac{dF}{dG_2}$ = b$\bar{c}$ + $\bar{b}$c, $\frac{dF}{dG_3}$ = $\bar{a}$bc + a$\bar{b}$ + a$\bar{c}$, $\frac{dF}{dG_4}$ = $\bar{a}$b$\bar{c}$ + a$\bar{b}$ + ac.

Now as per the second phase of the proposed testing method, the Boolean generator of the circuit is computed by the statement B_{Gen}, where B_{Gen} = $(\frac{dF}{dG_0})$AND$(\frac{dF}{dG_1})$AND$(\frac{dF}{dG_2})$ AND$(\frac{dF}{dG_3})$AND$(\frac{dF}{dG_4})$ = (bc)AND(c) AND(b$\bar{c}$ + $\bar{b}$c)AND($\bar{a}$bc + a$\bar{b}$ + a$\bar{c}$) AND($\bar{a}$b$\bar{c}$ + ab + ac) = 0. As B_{Gen} is 0, we have to find out the B^1_{Gen} and B^2_{Gen} to calculate the final form of the Boolean generator.

For this example, as gate count is N = 5, we need 5 iterations to generate the resultant Boolean generator of the circuit.

First Iteration: Let us assume B^1_{Gen} = $\frac{dF}{dG_0}$ = bc.

Second Iteration: Here B^1_{Gen} is upgraded as follows: B^1_{Gen} = B^1_{Gen} AND$(\frac{dF}{dG_1})$ = (bc)AND(c) = bc.

Third Iteration: Similarly, $B^1_{Gen} = B^1_{Gen} AND(\frac{dF}{dG_2}) = (bc) AND(b\bar{c} + \bar{b}c) = 0$.
That means the term $(b\bar{c} + \bar{b}c)$ converts the term B^1_{Gen} to 0 and therefore, the $B^2_{Gen}(0)$ is computed and B^1_{Gen} will not be upgraded. Now, $B^2_{Gen}(0) = (b\bar{c} + \bar{b}c)$.

Fourth Iteration: $B^1_{Gen} = B^1_{Gen} AND(\frac{dF}{dG_3}) = (bc) AND(\bar{a}b\bar{c} + a\bar{b} + u\bar{c}) = \bar{a}b\bar{c}$.

Fifth Iteration: $B^1_{Gen} = B^1_{Gen} AND(\frac{dF}{dG_4}) = (\bar{a}bc) AND(\bar{a}b\bar{c} + a\bar{b} + ac) = 0$.
Once again as per the third iteration, the $B^2_{Gen}(1)$ is computed and the function B^1_{Gen} will not be modified. Now, $B^2_{Gen}(1) = (\bar{a}b\bar{c} + a\bar{b} + ac)$.

Now, we need to compare B^1_{Gen} with $B^2_{Gen}(0)$ and $B^2_{Gen}(1)$ once again to upgrade the B^1_{Gen}. At First, the expression $B^1_{Gen}(\bar{a}bc)$ is compared with both the terms of $B^2_{Gen}(0)$. The term $(\bar{a}bc)$ of B^1_{Gen} contains a single literal matching with both the terms of $B^2_{Gen}(0)$ and hence, $B^1_{Gen}(0) = (\bar{a}bc + b\bar{c})$ and $B^1_{Gen}(1) = (\bar{a}bc + \bar{b}c)$.

Now, we need to compare both the B^1_{Gen} with the $B^2_{Gen}(1)$ to determine the updated value of $B^1_{Gen}(0)$ and $B^1_{Gen}(1)$. The $B^1_{Gen}(0)$ is compared with the $B^2_{Gen}(1)$ and the term $(\bar{a}bc)$ of $B^1_{Gen}(0)$ and the term $(\bar{a}b\bar{c})$ of $B^2_{Gen}(1)$ has the highest literal matching. So, the $B^1_{Gen}(0)$ is upgraded to $B^1_{Gen}(0) = (\bar{a}bc + b\bar{c} + \bar{a}b\bar{c})$. Similarly, the $B^1_{Gen}(1)$ is upgraded to $B^1_{Gen}(1) = (\bar{a}bc + \bar{b}c + \bar{a}b\bar{c})$.

After the minimization, the $B^1_{Gen}(0) = (\bar{a}b + b\bar{c})$ and $B^1_{Gen}(1) = (\bar{a}b + \bar{b}c)$.

As both $B^1_{Gen}(0)$ and $B^1_{Gen}(1)$ contains similar number of literals, there will be only two generators to identify all the possible SMGF in the given circuit. generator(0) $= B^1_{Gen}(0) = (\bar{a}b + b\bar{c})$ and generator(1) $= B^1_{Gen}(1) = (\bar{a}b + \bar{b}c)$

Now as per the third phase, the test vector formation from the Boolean generator is explained as follows:
$Minterm_{generator(0)} = \bar{a}b(c + \bar{c}) + b\bar{c}(a + \bar{a}) = \bar{a}bc + \bar{a}b\bar{c} + ab\bar{c} + \bar{a}b\bar{c} = \bar{a}bc + \bar{a}b\bar{c} + ab\bar{c} = \{3, 2, 6\} = \{2, 3, 6\}$. The test set derived from the generator(0) is $\{2, 3, 6\}$. Similarly, $Minterm_{generator(1)} = \bar{a}b(c + \bar{c}) + \bar{b}c(a + \bar{a}) = \bar{a}bc + \bar{a}b\bar{c} + a\bar{b}c + \bar{a}\bar{b}c = \{3, 2, 5, 1\} = \{3, 2, 5, 1\}$.

The test set derived from the generator(1) is $\{1, 2, 3, 5\}$. As the generator(0) contains lesser number of test vectors, the generator(0) will be used to uncover the SMGF fault.

4.2 Boolean Based Testing Method2

In this method, the ESOP-based reversible circuit is considered for the detection of SMGF fault and also to diagnosis the detected fault. Let us assume that C_{test} is the testable circuit in which the test has to be performed. To test the C_{test} circuit, initially a fault-free ESOP-based circuit (C_{true}) is read from a given specification files (T_{spec}) and then the logical XOR is performed between C_{true} and C_{test}.

But the design of fault-free ESOP circuit from the T_{spec} creates problem because number of distinct ESOP circuits can be generated from the T_{spec} and due to this reason MPRM can be considered as the subclass of an ESOP expression. A fault-free ESOP circuit (C_{true}) can be identified from a number of distinct ESOP circuits by the help of some complex calculation. To solve the said problem, we have used the PPRM class from which only one circuit can be designed.

The proposed testing method is segmented into two stages. At first, fault detection is performed in the C_{test} and after that in second stage, identification of the detected fault in the C_{test} is performed and proper diagnosis is done in the C_{test} to transform the circuit from faulty to fault-free circuit.

4.2.1   Detection of SMGF

Here in this phase, a PPRM expression f_{true}^{PPRM} from the given circuit specification file T_{spec} is obtained for the circuit, C_{test}. After that the MPRM expression f_{test}^{MPRM} is derived from testable ESOP circuit C_{test}. Now, for each variable contained within MPRM expression (f_{test}^{MPRM}), the polarity of such variables is converted to positive form and thereby a FPRM form (f_{test}^{FPRM}) can be derived. Now, the L_{XOR} is computed as follows: $L_{XOR} = f_{true}^{PPRM} \oplus f_{test}^{FPRM}$, where $\oplus$ denotes the XOR operation. If L_{XOR} is zero that means fault is not detected in the C_{test} else SMGF is detected in the C_{test}.

4.2.2   Detection and Correction of SMGF

Here, both fault identification and correction of the specified fault in a given circuit. The output expression obtained from L_{XOR} is considered as the specification details of the corresponding missing gate and represented in the form of a Boolean expression and the variables contained within such expression L_{XOR} designates the control inputs for the gate Mg, where Mg denotes the missing gate in C_{test} circuit.

To make the circuit function correctly, if the identified missing gate (M_g) is attached to the input circuit C_{test} to convert it to a fault-free circuit.

Now, the proposed method of SMGF identification and medication in an ESOP based circuit has been discussed with supportive examples below.

Example 2 The benchmark ESOP circuit 4gt4 [22] is used here for testing the proposed methodology. The T_{spec} of 4gt4 is represented in Fig. 5a. The testable circuit, C_{test} for the given circuit 4gt4 is also shown in Fig. 5c. Now, the PPRM cover of the fault-free circuit is obtained from file T_{spec} employing [23]. In Fig. 5b, the corresponding PPRM cube can be observed and the derived PPRM expression is $f_{true}^{PPRM} = a \oplus bd \oplus bc \oplus bcd \oplus abd \oplus abc \oplus abcd$. Now, the obtained expression from the circuit C_{test} is $f_{test}^{MPRM} = a \oplus \bar{a}b\bar{c}\bar{d} \oplus \bar{a}b$ and changing the polarity of each literal in the expression f_{test}^{MPRM} to positive value is required to derive the corresponding FPRM expression (($f_{test}^{FPRM} = a \oplus bd \oplus bc \oplus bcd \oplus abd \oplus abc \oplus abcd$). Now L_{XOR} is computed as $L_{XOR} = f_{true}^{PPRM} \oplus f_{test}^{FPRM} = \emptyset$.

Hence, it can be confirmed that there an SMGF does not exist in the circuit C_{test} as L_{XOR} is null.

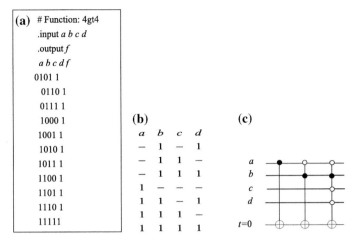

Fig. 5 Illustration of Example 2. **a** Input specification file of *4gt4* **b** Equivalent PPRM cube list of *4gt4* **c** Testable input ESOP circuit for function *4gt4*

Example 3 The benchmark circuit 4mod5 [22] is considered here once again to explain the proposed technique. The specification file T_{spec} for 4mod5 and C_{test} circuit are represented in Fig. 6a and b, respectively. Initially, $f^{PPRM}_{true} = 1 \oplus ad \oplus ab \oplus bc \oplus cd \oplus a \oplus b \oplus c \oplus d$ is obtained from T_{spec}. Then, MPRM expression ($f^{MPRM}_{test} = 1 \oplus a\bar{d} \oplus \bar{a}b \oplus \bar{b}c$) is derived from the circuit C_{test}. Now, an FPRM logic expression ($f^{FPRM}_{test} = 1 \oplus ad \oplus ab \oplus bc \oplus a \oplus b \oplus c$) is formed. The equivalent ESOP structure of derived f^{FPRM}_{test} and f^{PPRM}_{true} expressions are represented in Fig. 6c and d, respectively. Now, $L_{XOR} = f^{PPRM}_{true} \oplus f^{FPRM}_{test} = cd \oplus d = \bar{c}d$.

Hence, it is confirmed that a SMGF fault is present in C_{test} exists as L_{XOR} is equal to some Boolean value. As per the proposed method, the expression L_{XOR} represents the control lines of the gate M_g (as depicted in Fig. 6e) which is completely missing in the testable circuit. Thereafter, it can be described that the M_g is having a negative control and positive control at lines *c* and *d*, respectively.

The circuit can be made fault free if the corresponding missing gate (M_g) is adjoined to the input circuit C_{test} and the resultant original ESOP structure of *4mod5* is depicted in Fig. 6f.

5 Experimental Results

We have tested both of our approaches against different benchmark suites [22]. The results obtained from first testing approach is summarized in Table 1, where first three columns represent the circuit name, the number of lines (n) and the number of gates (N) present in a benchmark function. The Boolean generator is shown in

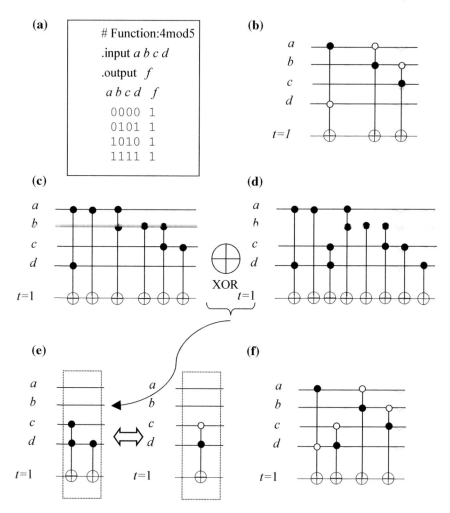

Fig. 6 Illustration of Example 3. **a** Specification file T_{spec} for circuit *4mod5* **b** Testable input circuit *4mod5* **c** f_{test}^{FPRM} function equivalent ESOP circuit **d** f_{test}^{FPRM} function equivalent ESOP circuit **e** Detected missing gate details **f** Fault free circuit of *4mod5*

column 4 and the set of test patterns produced from the generator are tabulated in column 5 of Table 1.

The second testing technique is also checked over several benchmark circuits and the effectiveness of RM form also has been verified for successful detection and localization of faults.

Table 1 Boolean generator for the detection of SMGF

Name of benchmark function	Number of lines (n)	Number of gates (N)	Boolean generator of the circuit	Derived test set from the boolean generator
rd32d1	4	4	$(ab + \bar{a}bc)$ or $(ab + a\bar{b}c)$	$\{6, 7, 12, 13, 14, 15\}$ or $\{10, 11, 12, 13, 14, 15\}$
xor5d1	5	4	$(ab\bar{c}\bar{d})$	$\{16, 17\}$
ham3tc	3	5	$(\bar{a}b + b\bar{c})$ or $(\bar{a}b + \bar{b}c)$	$\{2, 3, 6\}$ or $\{1, 2, 3, 5\}$
3_17tc	3	6	$(a\bar{b}c + ab\bar{c})$ or $(a\bar{b}c + \bar{a}b\bar{c})$	$\{5, 6\}$ or $\{5, 0\}$
mod5d1	5	8	$(abcd)$	$\{30, 31\}$
mod5d2	5	9	$(\bar{a}bcd)$	$\{6, 7\}$
4_49d3	4	12	$(abd + \bar{a}cd)$	$\{3, 7, 13, 15\}$
hwb4d1	4	17	$(a\bar{b}cd + \bar{c}d + ac\bar{d})$	$\{1, 5, 9, 10, 11, 13, 14\}$
2of5d1	6	15	$(\bar{a}c\bar{e} + a\bar{b}d\bar{e} + \bar{a}bd\bar{e} + \bar{a}b\bar{c}df)$	$\{8, 9, 12, 13, 20, 21, 23, 24, 25, 28, 29, 34, 35, 42\}$
mod5adders1	6	21	$(abc\bar{e}\bar{f} + ab\bar{c}d\bar{e} + ad\bar{e}f + \bar{b}c\bar{e}f)$	$\{0, 4, 32, 36, 40, 48, 49, 56, 60\}$

6 Conclusions

This work has presented two Boolean based approaches for testing of SMGF faults in reversible circuit. In the first method, a Boolean generator has developed for a reversible circuit and in later time, from this generator test vectors are constructed to a test a circuit. The second testing technique has mainly targeted to test a special class of reversible circuit known as ESOP designs. But, the first testing technique is very generic and can be employed over any type of circuits. In both the testing approaches, we have addressed SMGF only, but other types of faults like RGF, MMGF also can be tracked by following the same strategy as used to find SMGF. The presented techniques have successfully tested over a wide spectrum of benchmarks also.

References

1. Landauer R (1961) Irreversibility and Heat Generation in the Computing Process. IBM J Res Dev 5:183–191
2. Bennett CH (1973) Logical Reversibility of Computation. IBM J Res Dev 17:525–532

3. Nielsen M, Chuang I (2000) Quantum Computation and Quantum Information. Cambridge University Press, Cambridge
4. Wille R, Keszocze O, Hillmich S, Walter M, Ortiz AG (2016) Synthesis of approximate coders for on-chip interconnects using reversible logic. In: Design, automation and test in Europe
5. Rauchenecker A, Ostermann T, Wille R (2017) Exploiting reversible logic design for implementing adiabatic circuits. In: International conference mixed design of integrated circuits and systems
6. Ramasamy K, Tagare R, Perkins E, Perkowski M (2004) Fault localization in reversible circuits is easier than for classical circuits. In: Proceedings of international workshop on logic and synthesis
7. Chakraborty A (2005) Synthesis of reversible circuits for testing with universal test set and C-testability of reversible iterative logic arrays. In: Proceedings of VLSI design, pp 249–254
8. Rahaman H, Kole DK, Das DK, Bhattacharya BB (2011) Fault diagnosis for missing-gate fault (SMGF) model in reversible quantum circuits. Int J Comput Electr Eng (Elsevier) 37:475–485
9. Mondal J, Das DK, Kole DK, Rahaman H, Bhattacharya BB (2013) On designing testable reversible circuits using gate duplication. In: International symposium on VLSI design and test, pp 322–329
10. Kole DK, Rahaman H, Das DK, Bhattacharya BB (2010) Derivation of optimal test set for detection of multiple missing-gate faults in reversible circuits. In: Proceedings of Asian test symposium, pp 33–38
11. Mahammad SN, Hari SKS, Shroff S, Kamakoti V (2006) Constructing online testable circuits using reversible logic. In: International symposium on VLSI design and test, pp 373–383
12. Toffoli T (1980) Reversible computing. Technical memo MIT/LCS/TM-151, MIT Lab for Computer Science
13. Fredkin E, Toffoli T (1982) Conservative logic. Int J Theor Phys 21(3–4):219–253
14. Feynman RP (1996) Feynman lectures on computation. Perseus books
15. Fazel K, Thornton MA, Rice JE (2007) ESOP-based Toffoli gate cascade generation. In: Pacific rim conference on communications, computers and signal processing (PacRim). Victoria, Canada, pp 206–209
16. Zhang YZ, Rayner PJW (1984) Minimization of Reed-Muller polynomials with fixed polarity. IEE Proc Comput Digit Tech 131(5):177–186
17. Harking B (1990) Efficient algorithm for canonical Reed-Muller expansion of Boolean function. IEE Proc Comput Digit Tech 137(5):366–377
18. Hayes JP, Polian I, Becker B (2004) Testing for missing-gate faults inreversible circuits. In: Proceedings of Asian test symposium, pp 100–105
19. Perkowski M, Biamonte J, Lukac M (2005) Test generation and fault localization for quantum circuits. In: Proceedings of international symposium on multi-valued logic, pp 62–68
20. Polian I, Hayes JP, Fienn T, Becker B (2005) A family of logical fault models for reversible circuits. In: Proceedings of Asian test symposium, pp 422–427
21. Kohavi Z (1978) Switching and finite automata theory, 2nd edn. Tata McGraw- Hill, New York
22. Wille R, Grosse D, Teuber L, Dueck GW, Drechsler R (2008) Revlib: an online resources for reversible functions and reversible circuits. IEEE ISMVL 24:220–225
23. Bandyopadhyay C, Rahaman H (2014) Synthesis of ESOP-based reversible logic using positive polarity reed-muller form. In: IEEE emerging trends in computing and communication (ETCC-2014), Calcutta, India

Online Testable Efficient Latches for Molecular QCA Based on Reversible Logic

Debajyoty Banik

Abstract Quantum computer is very advanced technology in upcoming era and can capable to solve any complex problem with very fast computation power. The proposed models for quantum computation are quantum dot cellular automata (QCA). The molecular QCA has the tendency to high error rates. In case of molecular quantum dot cellular automata, the main objective of circuit design is the reduction of circuit area with wanted functional behavior. In this article, we propose the efficient design of online testable latches based on reversible logic for molecular QCA that is very much cost-effective respect to circuit area and other parameters. We use reversible gates having conservative property; i.e., capability for producing the equal number of 1s in output bits as input bits. So, conservative logic gates are subset of parity preserving reversible gate. Fault patterns of used conservative logic gate are analyzed for single stuck-at faults in molecular QCA circuit. Our proposed latches are able to examine single stuck-at fault, missing/additional QCA cell defect online, including permanent or transient fault in molecular QCA and efficient respect to circuit area. The designs of QCA layouts for various latches are presented and verified using QCA Designer and the Verilog HDL library of QCA devices is used to present HDLQ design tool.

1 Introduction

Molecular QCA is most demandable emerging nanotechnology having small shaped component, ultra low power consumption and high clock frequency [1, 2]. Molecular QCA cell can define the logic states, such as logic 0 or logic 1, by depending upon the electrons' location in it. Due to the tendency of high error rate in nanotechnology, it is necessary to check the error frequently in the circuit. And it is possible only when the circuit has online testable capability. Defect in molecular QCA can occur in various phases, such as synthesis, deposition, and runtime; missing/additional defects take

D. Banik (✉)
Indian Institute of Technology Patna, Patna, India
e-mail: debajyoty.pcs13@iitp.ac.in

© Springer Nature Singapore Pte Ltd. 2020
A. K. Singh et al. (eds.), *Design and Testing of Reversible Logic*, Lecture Notes
in Electrical Engineering 577, https://doi.org/10.1007/978-981-13-8821-7_11

place in synthesis and deposition phase [3]. Stuck-at fault and transient fault are more likely to take place in runtime for external unwanted energy or internal cell defect.

We introduce an efficient design of online testable latches for molecular QCA. The molecular QCA has various application of reversible computation. The reversible computation is possible only if the primary elements of the system are reversible gate. A reversible gate computes the function with bijective in manner. A reversible gate is $n \times n$ gate which has same number (n) of inputs and outputs with bijective mapping within input and output vectors. Landauer has proved that $k_B T ln2$ Joules energy is dissipated for every bit of information loss which is possible only irreversible logic [4]. Where, k_B is the Boltzmann constant and T represents the operating temperature. This energy dissipation Will not be occurred with reversible computation, firstly proposed by Bennett [5]. The previous exertion for applying reversible logic in molecular QCA is described in [6], in which reversible logic testing properties are described in a 1-D array of QCA cells.

The conservative reversible gates are used as basic component of the latches to provide online testability in the circuits and to reduce energy dissipation.

Conservative logic belongs to the logic family having a power to generate same number of 1s in the input and the output signals [7].

Definition 1: Bit conserving circuit is referred as conservative logic circuit. It must hold same number of 1s in input and output signals.

Definition 2: A circuit with is conservative in nature with having the reversibility property is known as conservative reversible logic circuit.

Definition 3: All high signals at the input lines of the conservative logic network are the cause to produce all high signals at the output lines, and all low signals at the input lines of the conservative logic network are the cause to produce all low signals at the output lines. Figure 1 shows the structure of very well-known reversible conservative Fredkin gate.

Here, the fault patterns for single stuck-at fault are analyzed. We get a clear idea from the fault patterns (for different type of faults) that when single fault will occur in the conservative logic gate it must mismatch the parity and specifically mismatch the number of 1s in output lines with input lines. Based on the above property we design online testable latches using conservative Fredkin gate. In existing paper, irreversible latches for molecular QCA are proposed without having online testable capability [8, 9]. Concurrently testable latches are proposed with decidedly less-cost efficient design which needs more molecular QCA cells to design [10]. So, the main contribution in this literature is to design various latches having online testable capability with cost-effective nature which needs few molecular QCA cells and minimal garbage lines.

Fig. 1 Conservative reversible Fredkin gate

This literature is arranged by the following way: Sect. 2 describes background work related to our work including the basics, cost metric of molecular QCA and also turn the attention to related work. Section 3 presents conservative reversible gate and online testing of molecular QCA. Various online testable reversible latches in molecular QCA framework are designed in Sect. 4. Implementation and Various simulation results of the proposed online testable latches in molecular QCA framework to verify our designs are shown in Sect. 5. Single stuck-at fault patterns are shown in tabular format and normal circuit simulation result is mentioned in graphical format to verify the designs in this section. Finally, Sect. 6 describes conclusions along with upcoming works.

2 Background

Various type of faults may occur in molecular QCA, like stuck-at fault [11], missing/additional cell defect which is more likely during synthesis and deposition phase for miss placement of the QCA cell [3], etc. It also has the tendency for permanent faults as well as transient faults caused by radiation, thermodynamic impact, and other effects, like the variation of power within excited and ground states is very small [12]. So, any type of single stuck-at fault, single cell defect (missing/additional) including permanent fault or transient fault detection are important. Our proposed designs covered all of such faults. Thus the proposed designs are significant.

2.1 Basics of molecular Quantum Cellular Automata Computing

QCA cell which is nothing but coupled dot system, is the basic unit of molecular quantum dot cellular automata nanotechnology circuit. Each cell contains four dots at the vertices of the square. Among four dots two electrons can quantum mechanically tunnel in this QCA cell. Due to electrostatic repulsion, the electrons are occupying diagonal positions, which is mentioned in Fig. 2a. Positions of electron pair in the QCA cell are caused to polarize the QCA cell either in logic 0 (P = −1) or in logic 1(P = +1), is mentioned in Fig. 2b.

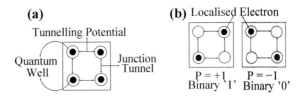

Fig. 2 QCA cell: **a** QCA cell formation. **b** Bi-polar QCA cell

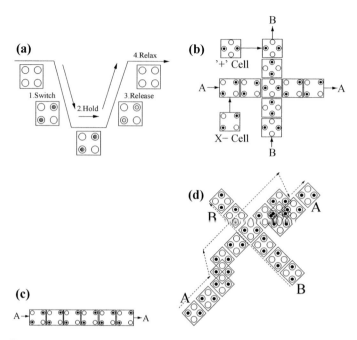

Fig. 3 **a** Clocking. **b** Coplanar wire-crossing. **c** Wire. **d** Multilayer wire-crossing

The four distinct periodic phases cascaded clocking is accomplished timing in QCA [4]. Figure 3a displays four distinct periodic phases are consist of switch, hold, release, and relax. Wire crossing in the molecular QCA is a remarkable achievement. The nature of the QCA cell makes wire-crossing more vulnerable. Undesirable crosstalk between wires may be introduced here. Wire-crossing determination is done either using different clock zones at different wires or using rotated QCA cells at one of the wires by coplanar approach. Considering 90° (X cell) and 45° (+ cell) structures form the coplanar wire crossing which is mentioned in Fig. 3b. The crossover can also be realized with multilayered wires. Instead of the coplanar strategy, multilayer strategy is more powerful [13]. Multilayer implementation of wire-crossing with same type of cells is mentioned in Fig. 3d.

Primary and the basic device of QCA computing is MV or majority voter, which is mentioned in Fig. 4. Property of majority voter is modeled as MV(A, B, C) = AB + BC + CA, which is a function to produce output from the majority of inputs A, B, and C. All other basic functions like OR, AND can be realized with majority voter. Two-input OR or a two-input AND gate can be realized through the fixation of a input cell at MV to p = +1 or p = −1, respectively. Realization of OR gate, AND gate is shown in Fig. 5.

INVERTER is another important logic device, which can be realized with many ways in QCA computing. Pictorial representation is mentioned in Fig. 6. Signal in QCA circuit may transfer through either inverter chain (Fig. 6c) or binary wire (Fig. 3c).

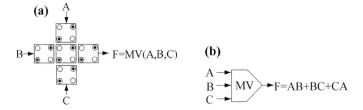

Fig. 4 Majority voter

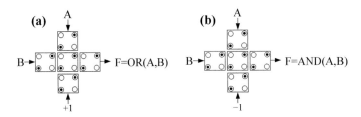

Fig. 5 **a** OR gate. **b** AND gate

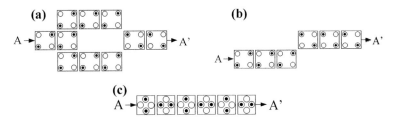

Fig. 6 **a** Inverter [type-1]. **b** Inverter [type-2]. **c** Inverter chain [type-3]

2.2 Cost Metric

Any circuit in molecular QCA framework must be designed with some cost-effective manner, which depends on number of QCA cell, majority voter, clock cycle is used to design the QCA circuit. We should reduce the cost metric to design efficient circuit.

Molecular QCA Cell

The QCA cell is primary element of the molecular quantum dot cellular automata. The QCA cell count should be minimum for reducing the circuit area. This is the most important cost metric because the main objective to design the circuit in any framework minimizes the area.

Majority Voter

In order to design the circuit with basic gates, majority voter is important to realize AND and OR gates. Minimum number of majority voters is considered as an efficient design to reduce the circuit area.

Clock Zone

The clock zone is another important cost metric because it estimates the time delay of the circuit. Thus, to make efficient circuit design we should minimize the delay or minimize the clock zone.

2.3 Related Work

Lent et al. proposed molecular QCA firstly in 1993 [14]. The first physically implemented molecular QCA produced Al islands and tunnel junctions. The experiment is done at 10mK [15]. As the circuit in molecular QCA having tendency to high error rate, so researchers were trying to concentrate on testing and the testing of molecular QCA first addressed in 2004 [3]. Here, the QCA devices defect characterization has been demonstrated, and also described how the difference between the testing of molecular QCA and the CMOS. QCA defects were modeled at molecular level for combinational circuits in [16]. It characterized faults in respect to single additional/missing cell defect on different QCA devices (i.e., INV, fan-out, MV, L-shape wire, cross wire, etc.). The test generation for molecular QCA described in [11]. For the QCA reversible logic circuit, single additional/missing cell defects were proposed in [6]. It concludes that reversible 1-D array must be C-testable. The fault-tolerant molecular QCA designs using the modular redundancy with the shifted operands, were shown in [17]. Considering various faults and wire delay, defect is modeled and SR latch is presented. The sequential circuits in QCA is presented based on the SR latch [8, 9]. The coplanar crossing in molecular QCA was presented and proved that wires with rotated cells can be considered as thermally more stable cells in [18]. The single stuck-at fault testing for combinational circuit is presented in [19]. Testing of reversible circuits is shown in [20–24] but they have not test circuit in of molecular framework. Though testable latches in QCA framework for missing/additional cell defect was done in [10], but it does not target to test stuck-at fault and it is not much efficient design respect to circuit area, time delay, etc.

Latches and flip-flops are the primary ingredients of various registers. Nowadays, researchers incorporate registers also for transliteration [25] to increase the processing speed. The automated transliteration and translation [26] are booming topics for now [27, 28]. Energy efficient fault- tolerant architecture is designed for wireless sensor networks [29]. Moreover, researchers introduced testable features into the sequential circuits which makes it more interesting [30–32]. In future, researchers may incorporate the proposed scheme in their applications [33] to improve efficiency.

3 Conservative Reversible Gate and Online Testing of Molecular QCA

The conservative reversible gate is a unique type of parity preserving logic which belongs to conservative reversible logic family. It has capability to form same number of 1s in output signals as the input signals. We shall discuss the conservative logic with case study of Fredkin gate. If anyone is interested to design the circuit with another conservative logic gate then they can also easily make the circuit online testable. The Fredkin gate is 3×3 conservative gate with reversible property. $I_v = \{A, B, C\}$ and $O_v = \{A, A'B\oplus AC, A'C\oplus AB\}$ having quantum cost 5; first proposed by Fredkin and Toffoli [7]. This is mentioned in Fig. 1. As the Fredkin gate belongs to the conservative reversible logic family. So, it is capable to produce equal number of 1's in input lines as output lines which can be cleared from behavioral analysis of the Fredkin gate which shown in Table 1. It should be noted that any type of single permanent or transient fault will cause to miss the conservative property as well as parity preserving property. Thus, depending upon the conservative property of the Fredkin gate, we can easily detect the fault online. Launder four-phase clocking scheme is most popular to design circuit in molecular QCA framework and our proposed designs are based on this clocking scheme. Figure 7 represents the Fredkin gate structure in molecular QCA.

The Fredkin gate is considered as basic building block of proposed latches based on the reversible logic in molecular QCA framework, having 6 FOs, 2 INVs, 6 MVs, 8 LSs, 5 CWs, and to implement it 233 QCA cells with 4 clock zone are needed.

In this work, we use Fredkin gate behalf of conservative logic gate. The Verilog library of the hardware description language notations for QCA layout is used to

Table 1 Behavioral study of the Fredkin gate

A	B	C	P	Q	R
Low	Low	Low	Low	Low	Low
Low	Low	High	Low	Low	High
Low	High	Low	Low	High	Low
Low	High	High	Low	High	High
High	Low	Low	High	Low	Low
High	Low	High	High	High	Low
High	High	Low	High	Low	High
High	High	High	High	High	High

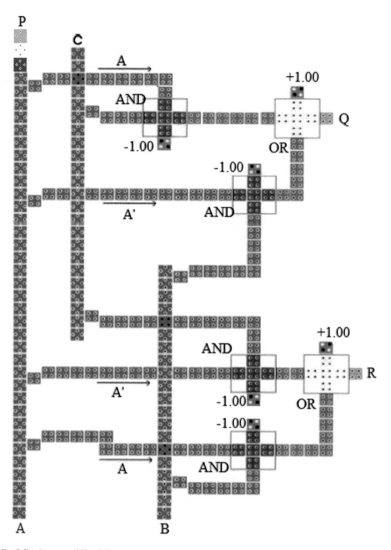

Fig. 7 QCA layout of Fredkin gate

show all possible single stuck-at fault and single cell defect (missing/additional) in majority voters (MJs), CWs, INVs, L-shape wires (LSs), FOs. The Verilog HDL library of molecular QCA devices is used as HDLQ design tool i.e., MV, FO, L-shape wire, Cross wire (CW), INV having ability to inject the faults [34]. Design HDLQ model for the Fredkin gate is mentioned in Fig. 8.

Verilog HDL simulator helps to simulate the proposed design and this is successfully done for the appearance of all the possible single stuck-at faults to determine corresponding erroneous outputs. 43 fault patterns were produced through the entire

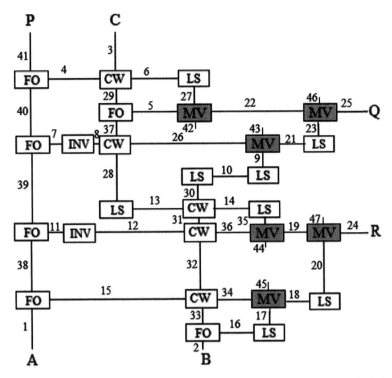

Fig. 8 Modeling for the Fredkin gate with QCA layout. INV describes inverter in QCA, FO illustrates QCA fanout device, MV describes QCA majority voter, LS shows QCA L-shape wire, and CW illustrates QCA cross wire

test for single stuck-at 0 and 45 fault patterns for single stuck-at 1. The fault patterns for single stuck-at 0 fault is mentioned in Tables 2 and 3, fault patterns for single stuck-at 1 fault is mentioned in Tables 4 and 5. In these tables, k represents the three-bit pattern of corresponding decimal value of roman k. s0 describes bit pattern 000 (decimal 0), a1 denotes bit pattern 001 (decimal 1), and so on. From the fault pattern table, it is clear that there must be mismatch in conservative nature if there exits any stuck-at fault. Conservative gate also mismatch in parity if there exist single cell defect (missing/additional) is occurred [10]. So, the fault patterns indicate the presence of fault, if any conservative discrepancy (i.e., number of 1s in input vector is unequal to number of 1s in output vector) is found between the input and output vectors. More generally parity mismatch helps to find the faults. So, any permanent or transient faults can be detected by parity mismatch property of conservative logic gate. Finally, it can be claimed that the Fredkin gate based on molecular QCA is online testable gate for its parity preserving property.

Table 2 Stuck-At 0 fault patterns in Fredkin gate (first 22 lines)

InV	EOV	Fault patterns																				
		1	2	3	4	5	6	7	8	9	10	11	12	13	14	15	16	17	18	19	20	21	22
0	0	0	0	0	0	0	0	0	0	0	0	0	0	0	0	0	0	0	0	0	0	0	0
i	i	i	i	0	i	i	i	i	i	i	i	i	0	0	0	i	i	i	i	0	i	i	i
ii	ii	ii	0	ii	ii	ii	ii	ii	0	0	0	ii	ii	ii	ii	ii	ii	ii	ii	ii	ii	0	ii
iii	iii	iii	i	ii	iii	iii	iii	iii	i	i	i	iii	ii	ii	ii	iii	iii	iii	iii	ii	iii	i	iii
iv	iv	0	iv	iv	iv	iv	iv	iv	iv	iv	iv	iv	iv	iv	iv	iv	iv	iv	iv	iv	iv	iv	iv
v	vi	i	vi	iv	iv	iv	iv	vi	vi	vi	vi	vii	vi	vi	vi	vi	vi	vi	vi	vi	vi	vi	iv
vi	v	ii	iv	v	v	v	v	vii	v	v	v	v	v	v	v	iv	iv	iv	iv	v	iv	v	v
vii	vii	iii	vi	v	v	v	v	vii	vii	vii	vii	vii	vii	vii	vii	vi	vi	vi	vi	vii	vi	vii	v

InV = Input vector, and EOV = Expected output vector

Table 3 Stuck-At 0 fault patterns in Fredkin gate (remaining lines)

InV	EOV	Fault patterns																				
		23	24	25	26	27	28	29	30	31	32	33	34	35	36	37	38	39	40	41	46	47
0	0	0	0	0	0	0	0	0	0	0	0	0	0	0	0	0	0	0	0	0	0	0
i	i	i	0	i	i	i	0	0	i	i	i	i	i	0	0	0	i	i	i	i	i	0
ii	ii	0	ii	0	0	ii	ii	ii	0	0	0	0	ii	ii	ii	ii	ii	ii	ii	ii	0	ii
iii	iii	i	ii	i	i	iii	ii	ii	i	i	i	i	iii	ii	ii	ii	iii	iii	iii	iii	i	ii
iv	iv	iv	iv	iv	iv	iv	iv	iv	iv	iv	iv	iv	iv	iv	iv	iv	0	0	0	0	iv	iv
v	vi	vi	vi	iv	vi	iv	vi	iv	vi	vi	vi	vi	vi	vi	vi	vi	i	0	0	ii	iv	vi
vi	v	v	iv	v	v	v	v	v	v	v	v	v	iv	v	v	v	iii	iii	i	i	v	iv
vii	vii	vii	vi	v	vii	v	vii	v	vii	vii	vii	vii	vi	vii	vii	vii	iii	iii	i	iii	v	vi

InV = Input vector, and EOV = Expected output vector

Table 4 Stuck-At 1 fault patterns in Fredkin gate (first 22 lines)

InV	EOV	Fault patterns																				
		1	2	3	4	5	6	7	8	9	10	11	12	13	14	15	16	17	18	19	20	21	22
0	0	iv	ii	i	0	0	0	0	0	ii	ii	0	0	i	i	0	0	0	i	i	i	ii	ii
i	i	vi	iii	i	iii	i	iii	i	i	iii	iii	0	i	i	i	i	i	i	i	i	i	iii	iii
ii	ii	v	ii	iii	ii	ii	ii	0	ii	ii	ii	ii	ii	iii	iii	iii	ii	ii	iii	iii	iii	ii	ii
iii	iii	vii	iii	iii	iii	iii	iii	i	iii	iii	iii	ii	iii	iii	iii	iii	iii	iii	iii	iii	iii	iii	iii
iv	iv	iv	v	vi	iv	vi	iv	iv	iv	iv	iv	iv	iv	iv	iv	iv	v	v	v	v	v	vi	vi
v	vi	vi	vii	vi	vi	vi	vi	vi	vi	vi	vi	vi	vii	vi	vi	vi	vii	vii	vii	vii	vii	vi	vi
vi	v	v	v	vii	v	vii	v	v	vii	v	v	v	v	v	v	v	v	v	v	v	v	vii	vii
vii	vii	vii	vii	vii	vii	vii	vii	vii	vii	vii	vii	vii	vii	vii	vii	vii	vii	vii	vii	vii	vii	vii	vii

InV = Input vector, and EOV = Expected output vector

Table 5 Stuck-At 1 fault patterns in Fredkin gate (remaining lines)

InV	EOV	Fault patterns																					
		23	24	25	26	27	28	29	30	31	32	33	34	35	36	37	38	39	40	41	42	43	44	45
0	0	ii	i	ii	0	0	i	i	ii	ii	ii	ii	0	i	0	i	iv	iv	iv	iv	0	ii	i	0
i	i	iii	i	iii	i	iii	i	i	iii	iii	iii	iii	i	i	i	i	vi	vii	vii	v	iii	iii	i	i
ii	ii	ii	iii	ii	ii	ii	iii	iii	ii	ii	ii	ii	iii	iii	ii	iii	iv	iv	vi	vi	ii	ii	iii	iii
iii	iii	iii	iii	iii	iii	iii	iii	iii	iii	iii	iii	iii	iii	iii	iii	iii	vi	vii	vii	vii	iii	iii	iii	iii
iv	iv	vi	v	vi	iv	iv	iv	vi	iv	iv	iv	iv	iv	iv	iv	iv	iv	iv	iv	iv	vi	iv	iv	v
v	vi	vi	vii	vi	vi	vi	vi	vi	vi	vi	vi	vi	vi	vi	vii	vi	vi	vi	vi	vi	vi	vi	vii	vii
vi	v	vii	v	vii	vii	v	v	vii	v	v	v	v	v	v	v	v	v	v	v	v	vii	vii	v	v
vii	vii	vii	vii	vii	vii	vii	vii	vii	vii	vii	vii	vii	vii	vii	vii	vii	vii	vii	vii	vii	vii	vii	vii	vii

InV = Input vector, and EOV = Expected output vector

4 Design of Online Testable Latches in molecular QCA Framework

In this part, we design efficient online testable latches based on minimum number of online testable conservative reversible Fredkin gate with minimal number of QCA cells, few majority voters, and significant clock zones.

Lemma *Any circuit based on online testable gates is also online testable.*   □

Proof Online testable capability denotes the power of fully fault coverage runtime. Here, we consider the single stuck-at fault. So, the test vector should cover all zeroes and all ones in the input lines and output lines. The fault cover achieves with either to generate all combination of outputs with some characteristic or adding extra line which defines the correctness of the circuit. Now, if the testable advantage achieves with the first characteristic then there is no problem to test each block, because it can be tested with comparing specified characteristic for every combination of input which is applied from preceding block. If the testable advantage achieves with the mentioned second characteristic then each block must generate error detection signal for each combination of provided input to that block. Finally, after getting green signals from every block it can easily confirm to have no fault in the circuit.

Lemma *Proposed designs of sequential circuit based on multilayer approach are online testable.*   □

Proof Multilayer based Fredkin gate in molecular QCA is online testable, proved in section III. According to Lemma 1, circuit based on Fredkin gate is online testable. As, it is sequential circuit, its input depends on output of previous state. According to conservative principle, generate output must have same number of 1's as applied 1's in input lines in fault-free case. Thus, after comparing equality in generated number of 1's at output with applied number of 1's including previous state signal to input of the Fredkin gate, and consider $\overline{Q}$ as extra line which must be opposite of Q, we can claimed that the generated signal is error free and the circuit is fault free.

4.1 D Latch (Positive Level Triggered)

Equation of the level triggered D latch (positive):

$$Q_{n+1} = (D \wedge E) \vee (\overline{E} \wedge Q_n)$$

where, D and E are input bit and enable line, respectively. Q_n and Q_{n+1} are primary outputs of the present state and the next state, respectively. As per the level triggered D latch (positive) characteristic equation, it is cleared the straightaway reflection of D passes to the output signal when E is high, as $Q_{n+1} = D$. Positive level triggered D latch stays identical in its past state when the enable signal is low; as $Q_{n+1} = Q_n$. Figure 9 shows the D latch (positive level triggered). Though fan-out is restricted at reversible circuit whereas fan-out is acceptable in molecular QCA. Thus, Fig. 9 is valid for low power QCA.

   As Fredkin gate has capability to detect fault online for its parity preserving characteristic, so the proposed D latch with the Fredkin gate also has the capability to detect fault online. To test the D latch, it is necessary to considered Q_n, E and D signals for input vector and G1, G2, Q_n as output vector; if parity mismatch is found within input and output vectors then it can be claimed that there is fault in the circuit. For testing the D latch (positive level triggered) with output Q and $\overline{Q}$, we only need to check same criteria and we should check additional $\overline{Q}$ bit which must be inverse of Q line if there exists no fault.

4.2 D Latch (Negative Level Triggered)

It will directly transfer the input signal D to Q line, when value of E is low; else estate will remain unchanged as before. The negative level triggered D latch can be modeled as,

$$Q_{n+1} = (D \wedge \overline{E}) \vee (E \wedge Q_n)$$

Fig. 9 Positive level triggered D latch with output Q and $\overline{Q}$

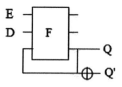

From the characterize equation, it is clear that D latch (negative level triggered) is mapped to Fredkin gate because of its MUXing capability, is mentioned in Fig. 10.

The proposed negative level triggered D latch also has capability to detect fault online. For testing, it is necessary to consider Q_n, E, and D signals for input vector and G1, G2, and Q signals as output vector. If parity mismatch is found within input and output vectors, then it can be claim that there is fault in the circuit because of its parity preserving property. For testing the D latch (negative level triggered) with output Q and $\overline{Q}$, only need to check same criteria and we should check additional $\overline{Q}$ bit which must inverse of Q bit for no fault.

4.3   T Latch

In this section, we design online testable T latch in molecular QCA framework. T latch is explained bellow:

$$Q_{n+1} = (T \oplus Q_n)E + \overline{E} Q_n$$

According to characterize equation of T latch, it is mapped to the conservative Fredkin gate. We design the online testable T latch with two Fredkin gates only. Our proposed design of online testable T latch with output lines Q and $\overline{Q}$ is mentioned in Fig. 11. To test the T latch online, it is necessary to consider Q_n, $\overline{Q}_n$, T, and E signals as input vector and G1, G2, G3, Q_{n+1}, and $\overline{Q}_{n+1}$ signals as output vector. Now if there is any mismatch in parity between the input and output vectors then this will be considered as faulty circuit.

4.4   JK Latch

In this section, online testable JK latch in QCA framework with two Fredkin gate is presented. The JK latch can be described as,

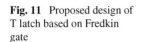

Fig. 10 Negative level triggered D latch

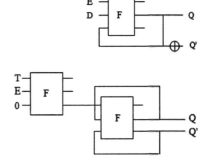

Fig. 11 Proposed design of T latch based on Fredkin gate

$$Q_{n+1} = (J\overline{Q}_n + \overline{K}Q_n)E + \overline{E}Q_n$$

From the characterize equation, it is clear that the JK latch can be realized using D latch or SR latch, but by this methodology circuit complexity will high, so alternative circuit diagram for JK latch is mentioned in Fig. 12.

The proposed JK latch based on the Fredkin gate also has capability to detect fault online. To test the JK latch, we should considered J, $\overline{K}$, E, and Q_n signals as input vector and G1, G2, G3, Q_{n+1} signals as output vector; if parity mismatch is found within input and output vectors, then it can be claimed that there is fault in the JK latch. For testing the JK latch with output Q and $\overline{Q}$, only need to check same criteria and we should check additional $\overline{Q}$ bit which must opposite of Q bit in fault-free case.

4.5   SR Latch

In this section, we propose the design of online testable SR latch in molecular QCA framework based on Fredkin gate. The characteristic equation of SR latch can be established as,

$$Q_{n+1} = (S + \overline{R}Q_n)E + \overline{E}Q_n$$

The characteristic equation can be mapped to D latch (positive level triggered) as $DE + \overline{E}Q_n$, where $D = (S + \overline{R}Q_n)$, and both of the equations can be realized using Fredkin gate. Our architecture for the testable SR latch is mentioned in Fig. 13.

Proposed SR latch based on the conservative Fredkin gate also have capability to detect fault online. To test the SR latch, we should consider all input bits including Q_n as input vector, and G1, G2, G3, G4, G5, G6, Q as output vector. If parity mismatch is found within input and output vectors, then it can be claimed to have fault in the SR latch. For testing the SR latch with output Q and $\overline{Q}$, only need to check same criteria and we should check additional $\overline{Q}$ bit which must opposite of Q bit in fault-free case.

Fig. 12 JK latch based on Fredkin gate

Fig. 13 SR latch based on Fredkin gate

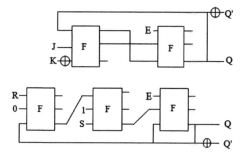

5 Implementation to Functionally Verify The Proposed Multilayer Latches in Molecular QCA Framework

Verification of every design is done by using the QCADesigner version 2.0.3 [35]. Following parameters are used for the bistable approximation, cell size $= 18$ nm, radius of effect $= 41$ nm, convergence tolerance $= 0.001000$, number of samples $= 182\ 800$, clock amplitude factor $= 2.000$, clock high $= 9.8\ e^{-22}$, clock low $= 3.8\ e^{-23}$, relative permittivity $= 12.9$, maximum iterations per sample $= 1000$, and layer separation $= 11.5000$ nm. In the QCA layouts, each multilayer Fredkin gate will produce output by one clock cycle delay since Fredkin gate is designed with four clock zones, as shown in Fig. 7. The simulated waveform of the multilayer conservative Fredkin gate is mentioned in Fig. 14 by which it verifies the functionality with one delay of the circuit. The applied inputs to multilayer Fredkin gate at clock zone 0 and the produced outputs from Fredkin gate at clock zone 4. The simulation result is exactly same as truth table of the Fredkin gate. It verifies the correctness of the proposed multilayer implementation of Fredkin gate. The important fact is that all the designs are practical as well as usable, since it was verified using the QCADesigner simulator that generated signals are proper without degradation. To perform correctly, all signals should appear simultaneously at the majority gate [36]. All designs follow this characteristic.

5.1 Multilayer QCA Layout and Simulation of the Proposed Latches

All the latches are implemented with proposed multilayer Fredkin gate. The QCA layout of the triple layer D latch (positive level triggered) is mentioned in Fig. 15 and it is verified using QCADesiger version 2.0.3 which is same as expected truth table. To better understanding of conservative property, in Fig. 15 garbage outputs are managed with proper clock zone. G1 and G2 represent the garbage output. To verify the design, the simulated results of the proposed triple layer D latch (positive level triggered) are mentioned in Fig. 16. Here, arrows are used to verify the functionality of the proposed triple layer D latch (positive level triggered). Arrow A, B, and C show that when E is high then D will be reflected in Q_{n+1}; like arrow A and arrow C indicate that when $E = 1$ and $D = 0$ then $Q_n = 0$. And since $E = 0$ in the next cycle, Q_{n+1} will maintain its previous value as 0 (mentioned with lines D and F), Arrow B indicate that when $E = 1$ and $D = 1$ then $Q = 1$. And since $E = 0$ in the next cycle, Q_{n+1} will maintain its previous value of Q_n as 1 (mentioned with lines E). To design the triple layer D latch (positive level triggered), 292 QCA cells, $0.44\ \mu m^2$ area and four clock zones are needed. All the generated output will be delayed by one clock cycle as Fredkin gate has one clock cycle delay. The Table 6 which is compact form of Fig. 16, summaries the working functionality of proposed triple layer D latch (positive level triggered).

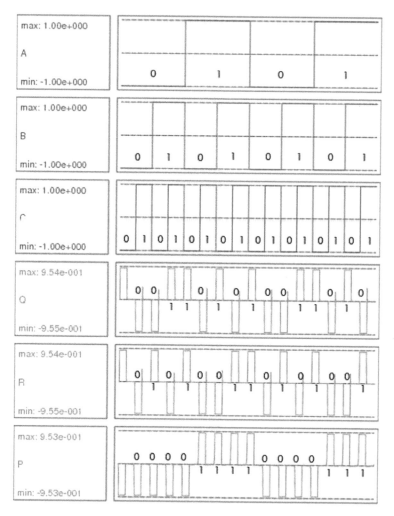

Fig. 14 Simulated results of multilayer Fredkin gate

Figure 17 shows multilayer representation of negative level triggered latch which is just opposite of positive level triggered latch. According to the characterize equation of negative level triggered latch, when E is low then D will be reflected in output Q_{n+1} and when E is high then the output Q_{n+1} will remain same as Q_n. To better understanding of conservative property, in Fig. 17 garbage outputs are managed with proper clock zone. G1 and G2 represent the garbage output. To verify the design, the simulated results of the proposed D latch (negative level triggered) are mentioned in Fig. 18. The trip of the arrows A and C indicate that when E = 0 and D = 0 then $Q_n = 0$. Since enable (E) = 1 in the next cycle, Q_{n+1} will maintain its previous value as 0 (mentioned with lines D and F), The trip of the arrow B indicates that when both

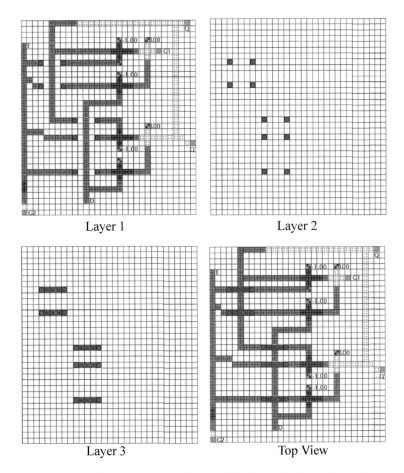

Fig. 15 Grid representation of proposed triple layer D latch (positive level triggered)

enable and the input line D are low then Q is based on previous state (i.e., high). And since E = 1 in the next cycle, Q_{n+1} will maintain its previous value of Q_n as 1 (mentioned with lines E). To design the triple layer D latch (negative level triggered), 272 QCA cells, $0.40\,\mu m^2$ area and four clock zones are needed. All generated outputs are delayed by a clock cycle. Table 7 which is compact form of Fig. 18, summaries the working functionality of proposed triple layer D latch (positive level triggered).

Figure 19 presents the top view representation of QCA layout and Fig. 20 show the simulation results of the triple layer T latch. To design the triple layer T latch, 628 QCA cells, $0.92\,\mu m^2$ area and eight clock zones are needed. Thus, after two clock cycles the correct output is generated. Table 8, summaries the working functionality of proposed triple layer T latch. The trip of arrow A represents toggling of Q_n at Q_{n+1} as when E = 1, T = 1, and $Q_n = 1$ then $Q_{n+1} = 0$. The Trip of the arrow B

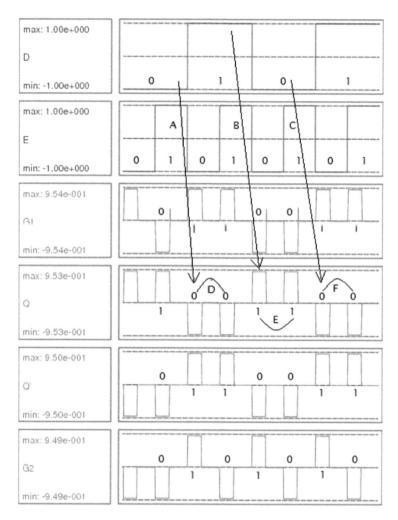

Fig. 16 Simulated output of the triple layer D latch (positive level triggered)

Table 6 Truth table of the D latch (positive level triggered)

Arrow	Input	Output (after one clock cycle)
A	E = High D = Low Q_n = High	Q_{n+1} = Low
B	E = High D = High Q_n = Low	Q_{n+1} = High
C	E = High D = Low Q_n = High	Q_{n+1} = Low
D	E = Low D = High Q_n = Low	Q_{n+1} = Low
E	E = Low D = Low Q_n = High	Q_{n+1} = High

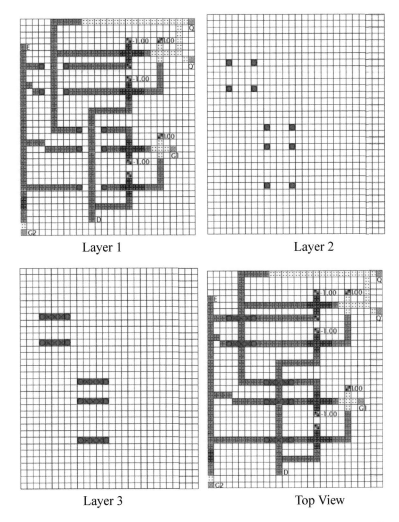

Fig. 17 Grid representation of proposed triple layer D latch (negative level triggered) with output Q based on multilayer Fredkin gate

represents when $E = 0$, $T = 0$, and $Q_n = 1$, then $Q_{n+1} = 1$. The trip of the arrow C represents when $E = 1$, $T = 0$, and $Q_n = 0$, then $Q_{n+1} = 0$.

Top view of the triple layer JK latch is represented in Fig. 21. The simulated results are shown in Fig. 22. To design the triple layer JK latch, 780 QCA cells, $1.50\,\mu m^2$ area and eight clock zones are needed, as two Fredkin gates are cascaded in series. Thus, the correct output will be generated after delay of two clock cycles. Table 9 which is compact form of Fig. 22, summaries the working functionality of proposed triple layer JK latch. The trip of arrow A represents when $E = 1$, $J = 0$, $K = 1$, and $Q_n = 0$, then $Q_{n+1} = 0$. The trip of arrow B represents when $E = 1$, $J = 1$, $K = 0$

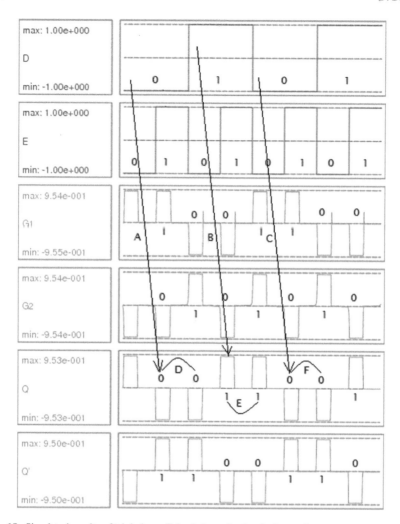

Fig. 18 Simulated results of triple layer D latch (negative level triggered)

Table 7 Truth table of the D latch (negative level triggered)

Arrow	Input	Output (after one clock cycle)
A	E = Low D = Low Q_n = High	Q_{n+1} = Low
B	E = Low D = High Q_n = Low	Q_{n+1} = High
C	E = Low D = Low Q_n = High	Q_{n+1} = Low
D	E = High D = Low Q_n = Low	Q_{n+1} = Low
E	E = High D = High Q_n = High	Q_{n+1} = High

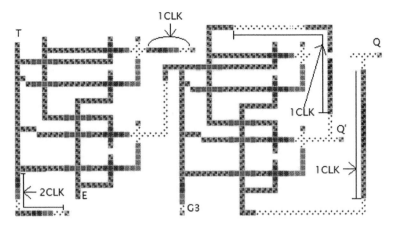

Fig. 19 Top view QCA layout of proposed triple layer T latch with output Q based on multilayer Fredkin gate

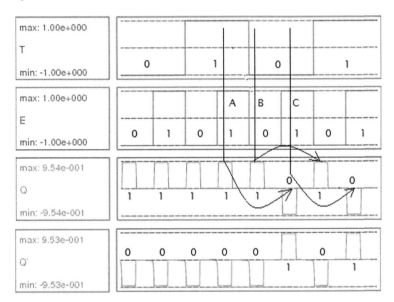

Fig. 20 Simulated results of triple layer T latch

Table 8 Truth table of the T latch

Arrow	Input	Output (after two clock cycles)
A	E = High D = High Q_n = High	Q_{n+1} = Low
B	E = Low D = Low Q_n = High	Q_{n+1} = High
C	E = High D = Low Q_n = Low	Q_{n+1} = Low

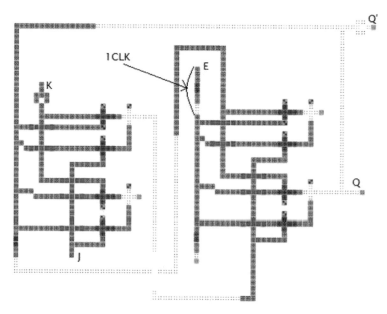

Fig. 21 Top view QCA layout of proposed triple layer JK latch with output Q using multilayer Fredkin gate

and $Q_n = 0$ then $Q_{n+1} = 1$. The trip of arrow C represents when $E = 1$, $J = 1$, $K = 1$, and $Q_n = 1$ then $Q_{n+1} = 0$. The trip of arrow D represents when $E = 1$, $J = 0$, $K = 0$, and $Q_n = 0$ then $Q_{n+1} = 0$. The trip of arrow E represents when $E = 0$, $J = 1$, $K = 0$, and $Q_n = 0$, then $Q_{n+1} = 0$.

Top view of QCA layout of the triple layer SR latch are presented in Fig. 23. The simulated results are shown in Fig. 24. To design the triple layer SR latch, 981 QCA cells, $2.17\,\mu m^2$ area and twelve clock zones are needed, as three Fredkin gates are cascaded in series. Thus, the correct output will be generated after delay of three clock cycles. Table 10 which is compact form of Fig. 24, summaries the working functionality of proposed triple layer SR latch. The trip of arrow A represents when $E = 1$, $S = 1$, $R = 0$ and $Q_n = 1$, then $Q_{n+1} = 1$. The trip of arrow B represents when $E = 1$, $S = 0$, $R = 1$ and $Q_n = 1$ then $Q_{n+1} = 0$. The trip of arrow C represents when $E = 0$, $S = 1$, $R = 1$ and $Q_n = 1$ then $Q_{n+1} = 1$. The trip of arrow D represents when $E = 1$, $S = 0$, $R = 0$, and $Q_n = 1$, then $Q_{n+1} = 1$.

Our multilayer latches are significant than the state-of-the-art latches in molecular QCA concerning with delays, Fredkin gates count, and the used number of QCA cells. A clock cycle delay is needed for each Fredkin gate in critical path. Comparison analysis of proposed multilayer latches in molecular QCA with existing one is mentioned in Table 11.

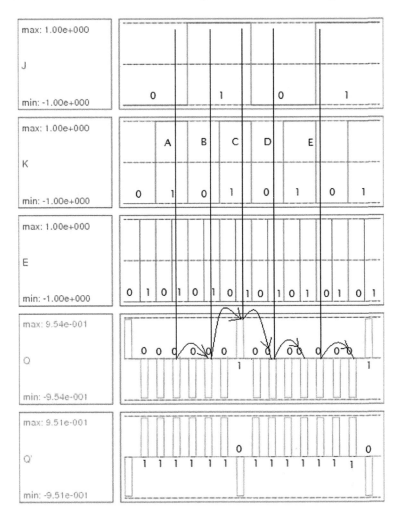

Fig. 22 Simulated results of triple layer JK latch

Table 9 Truth table of the JK latch

Arrow	Input	Output (after two clock cycles)
A	E = High J = Low K = High Q_n = Low	Q_{n+1} = Low
B	E = High J = High K = Low Q_n = Low	Q_{n+1} = High
C	E = High J = High K = High Q_n = High	Q_{n+1} = Low
D	E = High J = Low K = Low Q_n = Low	Q_{n+1} = Low
E	E = Low J = High K = Low Q_n = Low	Q_{n+1} = Low

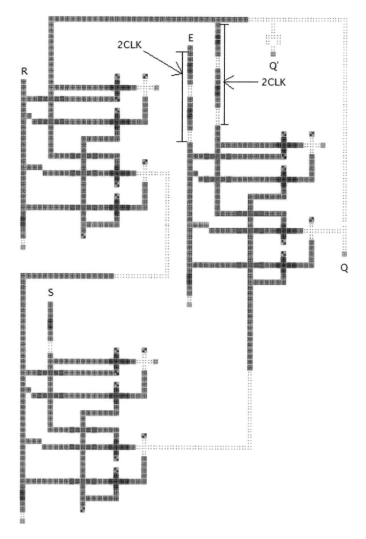

Fig. 23 Top view QCA layout of proposed triple layer SR latch with output Q based on multilayer Fredkin gate

6 Conclusions and Future Work

We propose online testable latches for molecular QCA using conservative logic (i.e., multilayer Fredkin gate). The proposed online testing approach concerning with parity preserving characteristics of the multilayer Fredkin gate without increasing the area of the circuit. It is beneficial for permanent fault as well as transient fault which can be detected by parity mismatch between inputs and outputs. $\overline{Q}$ may not be considered depending upon the prior information about circuit design, where

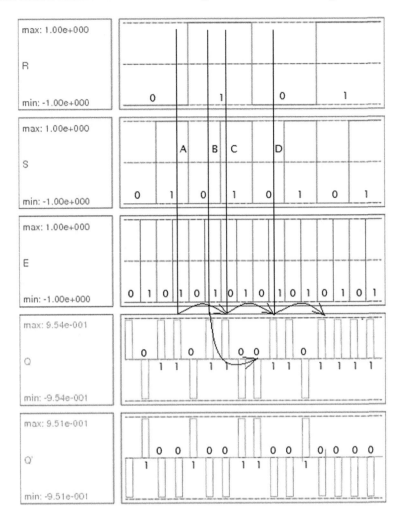

Fig. 24 Simulated results of triple layer SR latch

$\overline{Q}$ is introduced separately using NOT gate. For that case $\overline{Q}$ will be tested using knowledge of Q, as $\overline{Q}$ must be opposite of Q in fault-free case. The online testable designs for various multilayer latches, QCA layouts, and their simulation results are presented. The proposed methodology is applicable for the online detection of the single cell defect (missing/additional) model, single stuck-at fault model, or unidirectional faults. The proposed methodology is not appropriate to detect the bidirectional multiple faults, say expecting fault free output vector is {1110}, and due to bidirectional faults generated output vector is {1101} where parity of the input vector and output vector is same, thus fault is not detected. The input signals can be regenerated at output lines due to the reversible property. Thus, input vector

Table 10 Truth table of the SR latch

Arrow	Input	Output (after three clock cycles)
A	E = High S = High R = Low Q_n = High	Q_{n+1} = High
B	E = High S = Low R = High Q_n = High	Q_{n+1} = Low
C	E = Low S = High R = High Q_n = High	Q_{n+1} = High
D	E = High S = Low R = Low Q_n = High	Q_{n+1} = High

Table 11 Comparison analysis of proposed multilayer latches

Various latches	# of Fredkin gates in critical path		Required delays		# of QCA cells	
	In [10]	Proposed	In [10]	Proposed	In [10]	Proposed
D latch (positive level triggered) with output Q and Q'	2	1	2	1	598	292
D latch (negative level triggered) with output Q and Q'	None in literature	1	None in literature	1	None in literature	272
T latch with output Q and Q'	3	2	2	2	826	628
JK latch with output Q and Q'	4	2	4	2	1206	780
SR latch with output Q and Q'	4	3	4	3	1224	981

and output vector will not only preserve the parity, it must be exactly same. This characteristic can be implemented with our proposed approach to resolve the pitfall (detection of bidirectional multiple faults) of our proposed methodology in future, and also our proposed methodology can be implemented to design testable memory in molecular QCA framework. Finally, we can conclude that our proposed multilayer latches are implementation of online testable sequential circuit in molecular QCA framework with multilayer approach which is efficient respect to the area, delay, etc.

References

1. Tougaw P, Lent C (1994) Logical devices implemented using quantum cellular automata. J Appl Phys 75(3):1818–1825
2. Orlov AO, Amlani I, Bernstein GH, Lent CS, Snider GL (1997) Realization of a functional cell for quantum-dot cellular automata. Science 277:928–930
3. Tahoori MB, Huang J, Momenzadeh M, Lombardi F (2004) Testing of quantum cellular automata. IEEE Trans Nanotechnol 3(4):432–442
4. Landauer R (1961) Irreversibility and heat generation in the computational process. IBM J Res Dev 5:183–191
5. Bennett CH (1973) Logical reversibility of computation. IBM J Res Dev 17:525–532
6. Ma X, Huang J, Metra C, Lombardi F (2008) Reversible gates and testability of one dimensional arrays of molecular QCA. Springer J Electron Testing 24(1–3):297–311
7. Fredkin E, Toffoli T (1982) Conselvative logic. Int J Theor Phys 21:219–253

8. Momenzadeh M, Huang J, Lombardi F (2005) Defect characterization and tolerance of QCA sequential devices and circuits. In: Proceedings of the defect fault tolerance VLSI system, pp 199–207

9. Huang J, Momenzadeh M, Lombardi F (2007) Analysis of missing and additional cell defects in sequential quantum-dot cellular automata. Integr VLSI J 40(1), 503–515

10. Thapliyal H, Ranganathan N (2010) Reversible logic-based concurrently testable latches for molecular QCA. IEEE Trans Nanotech 9(1)

11. Gupta P, Jha NK, Lingappan L (2007) A test generation framework for quantum cellular automata circuits. IEEE Trans VLSI Syst 15(1)

12. Kummamuru RK, Orlov AO, Ramasubramaniam R, Lent CS, Bernstein GH, Snider GL (2003) Operation of a quantum-dot cellular automata (QCA), shift registers and analysis of errors. IEEE Trans Electron Devices 50(9):1906–1913

13. Schulhof G, Konrad W, Jullien G (2007) Simulation of random cell displacements in QCA. J Emerg Technol Comput Syst 3:1

14. Lent CS, Tougaw PD, Porod W, Bernstein GH (1993) Quantum cellular automata. Nanotech. 4:49–57

15. Orlov AO, Amlani I, Bernstein GH, Lent CS, Snider GL (1997) Realization of functional cell for QCA. Science 277, 928–930

16. Momenzadeh M, Ottavi M, Lombardi F (2005) Modeling QCA defects at molecular level in combinational circuits. In: Proceedings of the DFT VLSI System, Monterey, pp 208–216

17. Wei T, Wu K, Karri R, Orailoglu A (2005) Fault tolerant quantum cellular array (QCA) design using triple modular redundancy with shifted operands. In: Proceedings of the 2005 conference on Asia South Pacific design automation, Shanghai, China, pp 1192–1195

18. Bhanja S, Ottavi M, Pontarelli S, Lombardi F (2006) Novel designs for thermally robust coplanar crossing in QCA. In: Proceedings of the 2006 design, automation and test in Europe, Munich, Germany, pp 786–791

19. Sultana S, Imam SA, Radecka K (2006) Testing QCA modular logic. In: Proceedings of the 13th IEEE international conference on electronics circuits and systems, Nice, France, pp 700–703

20. Patel KN, Hayes JP, Markov IL (2004) Fault testing for reversible circuits. IEEE Trans Comput-Aided Design Integr Circuits Syst 23(8), 1220–1230 (2004)

21. Vasudevan DP, Lala PK, Parkerson JP (2006) Reversible-logic design with online testability. IEEE Trans Instrum Meas 55(2):406–414

22. Thapliyal H, Vinod AP (2010) Design of reversible sequential elements with feasibility of transistor implementation. In: Proceedings of the 2007 IEEE international symposium on circuits and systems, New Orleans, LA, pp 625–628

23. Chuang ML, Wang CY (2007) Synthesis of reversible sequential elements. In: Proceedings of the IEEE Asia south pacific design, automation in conference, Yokohama, Japan, pp 420–425

24. Rice JE (2006) A new look at reversible memory elements. In Proceedings of the international symposium on circuits and systems, Kos, Greece, pp 243–246

25. Banik D, Bhattacharyya P, Ekbal A (2016) Rule based hardware approach for machine transliteration: a first thought. In: 2016 Sixth international symposium on embedded computing and system design (ISED). IEEE

26. Banik D et al (2016) Can SMT and RBMT Improve each other's performance?-An experiment with English-Hindi translation. In: Proceedings of the 13th international conference on natural language processing

27. Sen S et al (2016) IITP English-Hindi machine translation system at wat 2016. In: Proceedings of the 3rd workshop on Asian translation (WAT2016)

28. Banik D, Ekbal A, Bhattacharyya P (2019) Machine learning based optimized pruning approach for decoding in statistical machine translation. IEEE Access 7:1736–1751

29. Mondal S, Banik D (2013) Energy efficient fault tolerant scheme for wireless sensor networks (EEFSWSN). IJCSET 3

30. Banik D, Mathew J, Rahamant H (2016) Testable reversible latch in molecular quantum dot cellular automata framework. In: India conference (INDICON), 2016 IEEE annual. IEEE

31. Banik D (2016) Design of low cost latches based on reversible quantum dot cellular automata. In: 2016 Sixth international symposium on embedded computing and system design (ISED). IEEE
32. Banik D (2016) Efficient methodology for testable reversible sequential circuit design. In: IEEE international conference on power electronics, intelligent control and energy systems (ICPEICES). IEEE
33. Banik D, Chowdhury RR (2016) Offline signature authentication: a back propagation-neural network approach. In: 2016 Sixth international symposium on embedded computing and system design (ISED). IEEE
34. Ottavi M, Schiano L, Lombardi F (2006) HDLQ: A HDL environment for QCA design. ACM J Emerg Technol 2(4):243–261
35. Walus K, Dysart T, Jullien GA, Budiman R (2004) QCADesigner: a rapid design and simulation tool for quantum-dot-cellular automata. Trans Nanotechnol 3(1):26–29
36. Kim K, Wu K, Karri R (2006) Quantum-dot cellular automata design guideline. IEICE Trans Fund E89-A(6), 1607–1614

Part IV
Applications to Emerging Technologies

An Efficient Nearest Neighbor Design for 2D Quantum Circuits

A. Bhattacharjee, C. Bandyopadhyay, B. Mondal, Robert Wille, Rolf Drechsler and H. Rahaman

Abstract In the last couple of years, synthesis of quantum circuits has received huge impetus among the research communities after the evolution of an efficient and powerful computational technology called "quantum computing". But physical implementation of these circuits considers the nearest neighbor qubit interaction as the desirable one otherwise a computational error can result. Realization of such an architecture in which qubit interacts only with its adjacent neighbors is termed as the *Nearest Neighbor* (NN) property. To attain such design architecture, SWAP gates plays a significant role of bringing the qubits to adjacent locations. But this in turn introduces design overhead so NN-based realization using limited number of SWAP gates has become significant. In order to explore this area, in this article, we introduced an efficient design technique for NN realization of quantum circuits in 2D architecture. The design algorithm has been partitioned into three phases of qubit selection, qubit placement and SWAP gate implementation. To verify the exactness of the stated design approach, its functionality has been evaluated over a wide set of benchmark function and subsequently witnessed an improvement on its cost metrics. By running our algorithm an overall improvement of about 17%, 3% against existing

A. Bhattacharjee (✉) · C. Bandyopadhyay · B. Mondal · H. Rahaman
Indian Institute of Engineering Science and Technology, Shibpur 711103, India
e-mail: anirbanbhattacharjee330@gmail.com

C. Bandyopadhyay
e-mail: chandanb@it.iiests.ac.in

B. Mondal
e-mail: bappa.arya@gmail.com

H. Rahaman
e-mail: rahaman_h@it.iiests.ac.in

R. Wille
Institute for Integrated Circuits, Johannes Kepler University Linz, 4040 Linz, Austria
e-mail: robert.wille@jku.at

R. Drechsler
Institute of Computer Science, University of Bremen & Cyber-Physical Systems, DFKI GmbH, 28358 Bremen, Germany
e-mail: drechsle@uni-bremen.de

© Springer Nature Singapore Pte Ltd. 2020
A. K. Singh et al. (eds.), *Design and Testing of Reversible Logic*, Lecture Notes in Electrical Engineering 577, https://doi.org/10.1007/978-981-13-8821-7_12

2D works and 35%, 22% against 1D works over SWAP gate count and quantum cost metrics have been recorded, respectively.

1 Introduction

Quantum computing, a new computational technology has shown promise in making the computation much easier by overcoming the limitations of conventional computing paradigm. Introduction of such computing technology not only finds solutions for some intractable problems but also solves them within a reasonable time bound.

Due to such computational facilities, quantum computing has left a remarkable footprint in the research community and thereby an effort has been evolved toward the establishment of quantum devices [1]. Additionally, quantum algorithms are required to be designed such that quantum computing architectures can be fabricated using quantum circuits. In quantum paradigm, quantum circuits are designed using a sequence of elementary quantum gates, representing quantum operators that manipulate quantum data information. To this end, qubits are considered as the basic quantum data units that acts similar to bits used in classical computing. However, these quantum units differ from the conventional data units in the way they exist. Unlike classical bits, qubits can be found to occur in multiple states simultaneously which can be represented as the linear superposition of the basis states.

The states of the quantum units are delicate and can easily be modified by environmental effects that can hamper the integrity of the quantum system. Therefore, an essential requirement toward attaining a practical and reliable quantum operation is by conducting fault-tolerant computation. For this purpose, quantum error correction codes turns out to be useful and thereby becomes acceptable [2]. However, application of these codes depends upon the nearest neighbor interaction of the qubits. Additionally, such a restriction in qubit interaction has also been considered as the limiting design constraint for the synthesis of certain quantum implementation technologies like ion trap [3], quantum dots [4], superconducting qubits [5] and nuclear magnetic resonance [6]. To make it more precise, the need toward realization of a nearest neighbor qubit interaction can be warranted due to the limitations of J-coupling force [7] required to enable multi-qubit operations (2-qubit or more) and can only be achieved effectively for adjacent neighboring qubits.

This design architecture can be obtained by making the quantum gates to act only on qubits located at adjacent positions, which can be realized via. SWAP gates. It can be made possible by exchanging the states of the qubits till the desired qubits become adjacent. Realization of NN architectures with the help of SWAP gates in turn causes an impact on the resultant architecture by enhancing the circuit depth and gate count. Therefore, NN optimization in terms of reduction in the number of SWAP gate requirements has become an essential design challenge such that the resulting overhead in the circuit can be checked. To address this, several articles related to efficient NN realization has been declared where the authors mainly considered SWAP gate reduction by following different design solutions. To realize NN design architecture, qubits need to be mapped onto the desired topological layout structure and the most commonly used structure is the qubit chain or 1D layout. Such an arrangement of qubits is referred to as the Linear Nearest Neighbor (LNN) architecture. Several contributions toward the development of LNN architecture have been made and detailed review of those works is presented next.

The importance of NN design networks along with the discussion of various design methodologies has been stated in the article [8]. In [9], a couple of efficient transformation schemes related to LNN synthesis is introduced that reduces the additional circuit and time complexity of the resultant structures. To produce cost-effective NN architectures, the authors of [10] suggested a transformation approach where mapping of circuits into lattice structure representations has been presented. The use of graph partitioning algorithm to form improved linear NN design architectures is stated in [11]. To improve the design further, two reordering techniques viz. global and local has been discussed in the work [12], where exact NN design solutions for each of the two reordering schemes are introduced so that the resultant solution becomes optimal. In pursuit of having an efficient NN realization, the authors of [13] presented a design methodology based on local reordering approach in which SWAP gate optimization is achieved by exploiting look-ahead strategies. To obtain efficient NN solutions for larger circuits, a heuristic approach using the look-ahead policy has been stated in the work [14]. Furthermore, an optimal linear NN architecture using global-based scheme has been obtained in [15], where an exact design algorithm based on A* approach is employed.

However, the stated research contributions discussed so far are based on LNN synthesis of quantum circuits where the qubits are arranged to form a chain like structure in which it can hardly have more than two adjacent neighbors to communicate. To maximize this nearest neighbor communication, the qubits need to be mapped to higher dimensional topological structures like 2D, 3D, and even on multidimensional layout. In other words, 2D organization of qubits facilitates to communicate with a maximum of four neighbors while such nearest neighbor communication can extend up to six in case of 3D structures. But such an increase in communication makes it difficult in controlling the qubits especially in 3D representations. Therefore, in this work, we have considered only on synthesis of 2D NN architectures. In the recent past, many research articles related to NN representations in 2D layouts have been reported and a few of them are stated below.

In the work [16], a mixed integer programming approach has been undertaken in which the design problem is formulated by mapping it into a 2D architecture and provides better design structures than 1D representation. An optimal 2D solution with respect to SWAP cost is derived in [17] whereby an exact design strategy has been employed. Despite the algorithm produces an optimal structure but it was found infeasible for large benchmark circuits as a result of extensive computational cost. In [18], the authors reduce the design overhead by arranging the qubits in such a manner that they are placed at suitable grid positions. To obtain a better NN representation, a heuristic design scheme has been undertaken in [19] in which a couple of grid selection approaches is employed followed by an efficient qubit placement strategy where the circuit's qubits are mapped based on their corresponding interaction count. To optimize the design further, the authors of [13] exploited a heuristic-based look-ahead design methodology that determines the appropriate movement of the qubits resulting in a significant reduction in SWAP cost. To provide an improved and scalable 2D NN representation, a better heuristic model based on generalization for a combined local and global reordering scheme has been investigated in the work [20].

Here, in this article we have employed a heuristic qubit mapping strategy to transform a quantum circuit to its corresponding 2D NN representation using less number of additional SWAP gates. Our design methodology has shown better performance over some of the existing research articles.

The remaining content of the article is formulated as follows. In Sect. 2, we discuss about the elementary quantum gates along with the nearest neighbor property. A detailed discussion of our design methodology has been presented in Sect. 3. An experimental evaluation of our approach against the related works has been summarized in Sect. 4. Finally, the article ends with concluding remarks in Sect. 5.

2 Background

In quantum technology, qubits are considered as the elementary quantum data units whose states undergo modification through execution of a sequence of quantum gates. Qubits share a similar characteristic of bits is that it can occur in one of the basis states like $|0\rangle$ and $|1\rangle$, which can be considered equivalent to 1 and 0 in conventional computing. In addition to these basis states, qubits can even occur in superimposed states of $|0\rangle$ and $|1\rangle$ that can be interpreted in the form of a state vector expression as

$$|\Psi\rangle = \alpha|0\rangle + \beta|1\rangle \tag{1}$$

where the notations α and β in the above expression denotes the complex numbers indicating the probability amplitudes of the corresponding basis states that satisfy the condition $\alpha2 + \beta2 = 1$. While measurement causes this state vector to degenerate into one of the basis states of $|0\rangle$ and $|1\rangle$.

The operations performed by the quantum gates on the qubits can be defined in the form of unitary matrices. In this context, the quantum functionality of an n-qubit quantum system can be realized via multiplication of distinct $2n \times 2n$ unitary matrices.

Definition 1 Quantum gates represent the elementary quantum operators that manipulate the qubits and when a collection of such gates are arranged over any group of circuit lines then the formed circuit is termed as quantum circuit.

The gates that have been most popularly used in the implementation of a quantum circuit are CNOT, NOT, and V/V+ and their corresponding symbolic representation have been represented in Table 1. These quantum gates form the constituent elements of the NCV gate library [21, 22] that help to map specific quantum into

Table 1 Symbolic representations of some quantum gates

Gates	Representation	Gates	Representation
NOT	⊕	Controlled-V	• V
CNOT	• ⊕		
V	V	Controlled-V†	• V†
V†	V†		

its corresponding gate level representations. In this chapter, our work is formulated using only the quantum gates from NCV library.

Despite the fact that quantum gates can be used to transform any function into a relevant circuit representation but there exist some physical limitations for the implementation of these circuits. To this end, realization of such circuits requires the quantum gates to interact only with its qubits physically placed at adjacent positions. In order to satisfy this design constraint, the process of implementing SWAP gates before any gate with non-neighboring qubits becomes essential that alters the positions of the qubits till the desired qubits become adjacent. Such a requirement for the physical design of quantum circuits is regarded as the Nearest Neighbor condition and it can be further described as follows:

Definition 2 Nearest Neighbor Cost of any 2-qubitgate g(c, t) can be interpreted as the distance separating the positions of its control (c) and target (t) qubits and this difference can be computed mathematically as

$$NNC(g) = |c - t| - 1 \qquad (2)$$

The above expression determines the nearest neighbor cost (NNC (g)) of an individual gate g and the combination of such costs of the respective gates produces the overall cost (NNC (QC)) of the corresponding quantum circuit QC. It can be represented mathematically as follows:

$$NNC(Q_C) = \sum NNC(g) \qquad (3)$$

This interpretation indicates that a given circuit holds the nearest neighbor condition provided either all the 2-qubit quantum gates act on adjacent qubits or it is having only 1-qubit gates. In other words, if the nearest neighbor expression represented above evaluates to zero then the resultant circuits are said to be NN-compliant.

Fig. 1 **a** Toffoli gate structure, **b** NCV realization of Fig. 1a

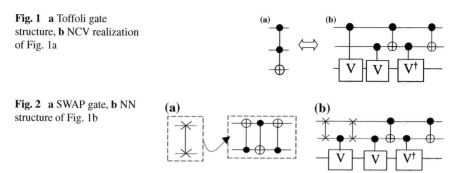

Fig. 2 **a** SWAP gate, **b** NN structure of Fig. 1b

Suppose, consider a Toffoli gate represented in Fig. 1a, whose corresponding *NCV* gate level realization obtained after decomposition is depicted in Fig. 1b. By inspecting Fig. 1b, it can be observed that the *NCV* realization of Toffoli gate does not hold the nearest neighbor condition as its overall NNC holds a positive (*NNC* $(Q_C) = 1$).

To make the circuit as NN-compliant, SWAP gates (diagrammatically represented in Fig. 2a) needs to be inserted before the first gate with nonadjacent interacting qubits which changes the qubit positions till they are placed adjacent. The resulting NN circuit post SWAP injection is depicted in Fig. 2b.

To get a clear understanding of the purpose of SWAP gate, another example has been considered in which transformation of a non-NN-complaint into its corresponding NN representation using SWAP gates is illustrated.

Example 1 Let's consider the circuit as depicted in Fig. 3a that fails to meet the nearest neighbor criteria as all the 2-qubit gates does not have their interacting qubits placed adjacent. SWAP gates are needed to transform the given circuit to its equivalent NN design. The resultant NN circuit is obtained after inserting ten SWAP gates as depicted in Fig. 3b.

The transformation process we have discussed so far is related to NN architectural design of 1D quantum circuit. It is also possible to obtain a much better NN representation for the corresponding circuit by projecting it in a 2D architectural format in which the qubits are mapped from a linear chain like structure to a two-dimensional

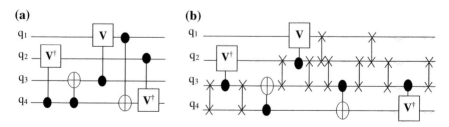

Fig. 3 **a** Quantum circuit with NNC = 5, **b** NN-compliant representation of Fig. 3a

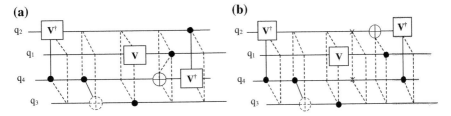

Fig. 4 **a** Orientation of circuit in Fig. 3a into 2D structure, **b** NN representation of Fig. 4a

grid structure. Such a mapping process produces an improved NN architecture over its linear counterpart by reducing the SWAP gate requirement. Now, we will discuss about the NN representation of quantum circuits in 2D architecture. The transformation process of quantum circuit into a two-dimensional topological layout can be carried out by mapping the qubits from linear format into a grid-like structure whereby the qubits are permitted to interact with four adjacent neighbors while only two nearest neighbor interaction can be allowed in 1D layout. Such an increment in nearest neighbor communication of qubits minimizes the number of nonadjacent gates and thereby leads to an efficient representation of NN structure with fewer SWAPS overhead in the circuit. Moreover, optimization of the resultant design structures depends on the selection of an appropriate 2D configuration grid used in the qubit mapping process.

Consider the following mapping of a five-qubit circuit into a 2D structure. There exist several possible solutions in which such a mapping process can be conducted depending upon the availability of grid configurations. Likewise, a five-qubit circuit can be represented in 2×3, 3×2, or 3×3.

Example 2: Considering the circuit shown in Fig. 3a, transformed into 2D structure in which a 2×2 configuration is chosen for arranging the qubits randomly as depicted in Fig. 4a. After inserting the necessary SWAP gates the resultant NN circuit realized in 2D topological layout is shown in Fig. 4b.

3 Proposed Approach

In this chapter, an improved heuristic qubit mapping scheme for the synthesis of NN circuits in 2D configuration has been described. This design workflow realizes an efficient NN architecture in which the circuit overhead is reduced by controlling SWAP gate implementation. For this purpose, our synthesis mechanisms are developed based on some heuristic policy that can be used for making some design-oriented decision and thereby formulated a unique qubit placement strategy for better mapping of qubits.

In this regard, our design workflow has been segmented into three phases namely qubit selection, followed by qubit placement and then SWAP gate insertion. For better realization, all the aforementioned stages have been explained with an illustrative example and for this any circuit specification such as the one shown in Fig. 5a is considered as the input on which all the desired operations pertaining to the synthesis approach are conducted.

Phase1: Qubit Selection Policy

The purpose of this phase is to arrange the qubits in a suitable manner on the 2D grid structure. To fulfill this objective, we have used some qubit preference metrics in which two metric tables viz. *time interaction* and *time costing* are computed by reading the gate level specifications of the given circuit.

Total interaction time of each individual qubit is estimated and recorded the results in a *time interaction* table. The values stored in this table constitute the interacting timestamps of the operating qubits and their corresponding overall interaction time is determined by aggregating the timestamps for each individual qubits in the given circuit. For the circuit shown in Fig. 5a, its *time interaction* table is estimated and represented in Table 2.

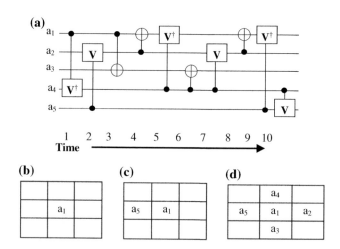

Fig. 5 **a** Input circuit, **b** a_1 inserted in the center of 3×3 grid, **c** a_5 placed on the left of a_1, **d** Qubits a_4, a_2, a_3 placed around a_1

Table 2 Qubit time interaction table

Qubits	Time instants	Total interaction
a_1	1, 3, 4, 5, 8, 9	30
a_2	2, 4, 7, 8	21
a_3	3, 6	9
a_4	1, 5, 6, 7, 10	29
a_5	2, 9, 10	21

Table 3 Qubit time costing table

Qubits	Time instants	Total interaction	Time instants	Total costing
a_1	1, 3, 4, 5, 8, 9	30	1, 3 ,5, 9	18
a_2	2, 4, 7, 8	21	2, 7	9
a_3	3, 6	9	3	3
a_4	1, 5, 6, 7, 10	29	1, 5, 7	13
a_5	2, 9, 10	21	2, 9	11

Table 4 Qubit preference table

Qubits	Total interaction	Total costing	Preference index
a_1	30	18	$18/30 = 0.6$
a_2	21	9	$9/21 = 0.42$
a_3	9	3	$3/9 = 0.33$
a_4	29	13	$13/29 = 0.44$
a_5	21	11	$11/21 = 0.52$

After resolving the qubit interaction metric discussed above then we work on the time-related cost of all the interacting qubits in the circuit. Computation of this cost parameter involves evaluation of the total costing time of the qubits and for this calculation purpose, the previous qubit interaction (*time interaction*) table determined is needed. Finally, the computation ends by recording the qubit costing results in a table called *time costing* (see Table 3). This parameter is determined by identifying the time steps in which a qubit of any 2-qubit gate is not interacting with an adjacent neighboring qubit using the respective qubit interaction time values and in this manner, such timestamp evaluation is carried out for all the individual qubits of the circuit provided.

After deriving these two tables, we merge the information contained in them and stored the combined result in a new table called qubit preference table is tabulated in Table 4. This preference table contains the qubit preference index evaluated using the ratio between the costing and interaction time values for all the qubit acting as inputs in a given circuit. Next, this preference table is sorted in decreasing order using the qubit index values computed earlier in which the qubit with largest indexing value is stored at the beginning of the table as displayed in Table 5. Now the index entries in the sorted preference table represent the qubit priority values used in the decision making purpose related to qubit placement as discussed in the next phase.

Phase2: Qubit Placement Policy
At the end of the previous workflow process, a preference table providing the sequence to be followed for qubits mapping on a 2D grid structure is considered as the input of this phase on which our mapping algorithm is executed as discussed next.

Table 5 Sorted qubit preference table

Qubits	Total interaction	Total costing	Preference index
a_1	30	18	0.6
a_5	21	11	0.523
a_4	29	13	0.448
a_2	21	9	0.428
a_3	9	3	0.33

The qubit mapping process works by picking the appropriate qubits from the preference table (PT) (see Table 5) based on their preference index values and thereby arranges the qubits on the chosen grid position. To be precise, the process starts by selecting the qubit with highest priority index from PT table and positioned the corresponding qubit at the center of the grid structure. After allocating the position of this qubit then a searching process is applied on the resultant grid to look for an empty cell having maximum number of adjacent vacant locations. After detection of such a desired position, the algorithm selects the next preferred qubit from preference table and places it in the identified location. If our search results does not generate a unique vacant cell then the qubits from PT table are placed in these locations by selecting them in the order of left, top, right and bottom with respect to the position of the last placed qubit depending on the space availability and carried out till a definite location is found. Following this algorithmic policy, ordered qubits from preference table are selected and settled on the 2D grid structure.

Now, this same mapping function is employed on the preference table generated in the previous phase to arrange the qubits on a proper 2D network. As a result, the qubit with high priority, a_1 is taken and placed at the center of the chosen grid (3 × 3) as shown in Fig. 5b. After placing qubit a_1, we initiate a search to look for cells surrounded with large number of adjacent empty locations on the corresponding grid. In this case, we have determined eight such possible cells surrounded by a maximum of two vacant cells and to solve this the remaining qubits from the preference table are selected orderly and placed by following the convention like left, top, right and bottom of the last qubit a_1 till a unique empty cell is identified. Hence, the next ordered qubit a_5 in PT is fetched and then placed at a location occurring to the left of a_1 as represented in Fig. 5c. In this manner, the remaining qubits viz. a_4, a_2, a_3 are mapped to the positions locating at the top, right, and bottom of a_1, respectively. The resultant grid structure obtained after placing all the qubits appear as shown in Fig. 5d.

Phase3: SWAP gate insertion

In the previous design phase, our mapping function has organized all the qubits on the given grid and thereby the circuit has been transformed into a 2D representation. Now, we consider the resultant 2D grid obtained in phase 2 and examine the gates with their qubits arranged on a grid and in this process if we identify a one with

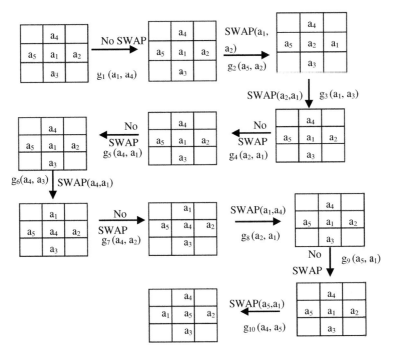

Fig. 6 Steps of SWAP gate insertion

nonadjacent qubits then a SWAP gate is applied to bring those qubits at adjacent positions.

Considering the circuit shown in Fig. 5a, execution of phase 2 generates the structure represented in Fig. 5d and now in this phase we work on this grid and apply SWAP gates at required positions so that an equivalent NN circuit is formed as depicted in Fig. 6. It can be noticed that overall of five SWAP gates are needed to transform the circuit into its corresponding NN form.

To have a better interpretation of the entire design mapping process, we have provided another illustrative example.

Example 3 Here, we have considered a benchmark function, 4gt11_84 as shown in Fig. 7, to describe the transformation mechanism used in our qubit mapping process is represented from Tables 6, 7, 8, 9, 10 (Figs. 8, 9).

4 Experimental Results

The qubit mapping algorithm has been developed using C and executed the function on a machine having Intel i5 processor with 4 GB RAM and 3.30 GHz clock. The performance analysis of our mapping scheme has been made by conducting experi-

Fig. 7 Input benchmark
circuit (4gt11_84)

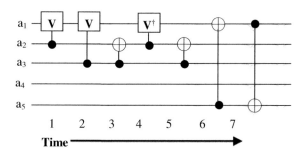

Table 6 Qubit time interaction table

Qubits	Time instants	Total interaction
a_1	1,2,4,6,7	20
a_2	1,3,4,5	13
a_3	2,3,5	10
a_4	–	–
a_5	6,7	13

Table 7 Qubit time costing table

Qubits	Time instants	Total interaction	Time instants	Total costing
a_1	1,2,4,6,7	20	2,6,7	15
a_2	1,3,4,5	13	0	0
a_3	2,3,5	10	0	0
a_4	–	–	–	–
a_5	6,7	13	6,7	13

Table 8 Qubit preference table

Qubits	Total interaction	Total costing	Preference index
a_1	20	15	$15/20 = 0.75$
a_2	13	0	$0/13 = 0$
a_3	10	0	$0/10 = 0$
a_4	–	–	–
a_5	13	13	$13/13 = 1$

Table 9 Sorted preference

Qubits	Total interaction	Total costing	Preference index
a_5	13	13	1
a_1	20	15	0.75
a_2	13	0	0
a_3	10	0	0
a_4	–	–	–

Table 10 Comparison results over medium size benchmark functions

Benchmark's name	Gate count	No. of qubits present	1D results from the work of [16]		2D Results from the work of [16]			2D Results from the work of [19]			Proposed work results		
			Quantum cost	SWAP count	Grid size	Quantum cost	SWAP count	Grid size	Quantum cost	SWAP count	Grid size	SWAP count	Quantum cost
3_17_13	14	3	17	4	2 × 2	20	6	2 × 2	17	3	2 × 2	5	19
4_49_17	32	4	45	12	2 × 2	45	13	–	–	–	2 × 2	10	42
aj-e11_165	60	4	96	36	2 × 3	84	24	3 × 2	82	22	2 × 3	18	78
decod24-v3_46	9	4	12	3	3 × 2	12	3	–	–	–	2 × 2	2	11
hwb4_52	23	4	33	10	2 × 2	32	9	2 × 2	32	9	2 × 2	7	30
rd32-v0_67	8	4	10	2	2 × 3	10	2	–	–	–	2 × 3	2	10
4gt11_84	7	5	8	1	2 × 3	9	2	2 × 3	9	2	2 × 3	2	9
4gt10-v1_81	36	5	56	20	3 × 2	52	16	3 × 2	51	15	2 × 3	14	50
4gt12-v1_89	53	5	88	35	3 × 2	72	19	2 × 4	71	18	3 × 2	20	73
4mod5-v1_23	24	5	33	9	2 × 3	35	11	3 × 2	31	7	2 × 3	8	32
4gt5_75	22	5	34	12	3 × 3	30	8	2 × 5	32	10	3 × 3	9	31
4gt4-v0_80	44	5	78	34	2 × 3	61	17	4 × 4	59	15	2 × 3	16	60
4mod7-v0_95	40	5	61	21	3 × 3	53	13	2 × 5	54	14	3 × 3	10	50
alu-v4_36	32	5	50	18	2 × 3	42	10	2 × 5	43	11	2 × 3	11	43
hwb5_55	109	5	172	63	3 × 2	154	45	2 × 7	158	49	3 × 2	38	147
QFT5	10	5	16	6	3 × 2	15	5	4 × 2	15	5	3 × 2	5	15

(continued)

Table 10 (continued)

Benchmark's name	Gate count	No. of qubits present	1D results from the work of [16]		2D Results from the work of [16]			2D Results from the work of [19]			Proposed work results		
			Quantum cost	SWAP count	Grid size	Quantum cost	SWAP count	Grid size	Quantum cost	SWAP count	Grid size	SWAP count	Quantum cost
hwb6_58	146	6	264	118	2 × 3	225	79	2 × 3	222	76	2 × 3	63	209
mod5adder_128	87	6	138	51	3 × 2	128	41	2 × 3	125	36	3 × 2	28	115
mod8-10_177	109	6	181	72	3 × 3	154	45	4 × 3	152	43	3 × 3	39	148
QFT6	15	6	27	12	2 × 3	21	6	3 × 2	22	7	2 × 3	5	20
rd53_135	78	7	144	66	5 × 2	117	39	2 × 7	118	40	3 × 3	29	107
ham7_104	87	7	155	68	3 × 3	135	48	2 × 7	132	45	3 × 3	38	125
QFT7	21	7	47	26	2 × 4	39	18	6 × 2	35	14	2 × 4	14	35
QFT8	28	8	61	33	4 × 2	46	18	4 × 2	51	23	4 × 2	18	46
QFT9	36	9	90	54	3 × 3	70	34	5 × 2	72	36	3 × 3	24	60
rd73_140	76	10	132	56	4 × 3	113	37	3 × 6	119	43	4 × 3	28	104
sys6-v0_144	62	10	121	59	4 × 4	93	31	–	–	–	4 × 3	30	92
QFT10	45	10	115	70	5 × 3	98	53	4 × 3	96	51	4 × 3	33	78
rd84_142	112	15	260	148	5 × 3	166	54	4 × 5	174	62	5 × 3	48	160
cnt3-5_180	125	16	252	127	3 × 6	194	69	4 × 4	209	84	3 × 6	54	179
Total			2796	971		2325	775		2179	740		628	2178

Fig. 8 Qubits arranged on
grid (2×3)

mental evaluations over a set of various benchmark functions taken from [23]. The
experimental data set has been summarized in two result tables Tables 10 and 11
respectively. The result sets for small and medium size benchmark specifications are
recorded in Table 10 while the other table contains the data of large size functions.
In each of these tables, two cost metric values namely quantum cost and SWAP cost
are evaluated for all the benchmark functions and compared the estimated results
against some existing 1D and 2D design approaches.

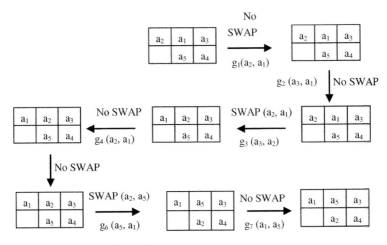

Fig. 9 Stages of SWAP insertion policy

Table 11 Comparison over higher size benchmarks

Benchmark's name	Initial QC incurred	Total lines present	Gate count	Grid size	QC in NN design	No. of SWAPs
rev_17	136	17	136	6×3	443	214
hm_20	73	20	73	5×4	142	69
ac_21_1	130	21	130	6×4	246	116
rev_18	153	18	153	5×4	374	221
ac_21_2	67	21	67	6×4	120	53
rev_19	171	19	171	4×5	427	256
hm_21	79	21	79	6×4	181	102
ac_21_3	42	22	42	6×4	93	51
hm_22	85	22	85	6×4	179	94

Table 10 contains a comparison analysis of our mapping algorithm with the results of the reported work [19] and against the best results of [16]. From this comparison, our proposed mapping approach has shown better performance over the reported works.

From the analysis of the result tables, an overall improvement of about 28.71% and 10.18% with respect to the previous 1D and 2D works has been estimated. In addition to this, a best case improvement of about 37.73% and 35.71% is noticed over the reported 2D articles [16] and [19] whereas an improvement of about 67.56% is attained in case of 1D [16] representation. Investigation of our result tables suggests that our design algorithm provides an improved resultant structure for majority of the benchmark functions. But for few circuits, our mapping approach has not provided a better NN realization than the reported ones.

5 Conclusion

In the present work, an improved heuristic qubit mapping scheme for NN realization of 2D quantum circuits is discussed. The design algorithm has been divided into three segments starting with qubit selection process followed by placement strategy and then SWAP insertion policy. To justify the functionality of our mapping scheme, it has been evaluated over a various set of benchmark circuits and has shown improved results. The computed values are compared with some of the existing 1D and 2D research articles. Based on our evaluation, it can be inferred that the proposed design mapping scheme attains a significant improvement around 35.32%, 22.10% over SWAP and quantum cost metric in 1D whereas an improvement of 17.09%, 3.28% over the same metrics is registered in case of 2D. In spite of having an improvement over the existing articles but the computed results may not be optimal because of implementing heuristic design policy in qubit mapping process and thereby in the future we work on optimizing the design structure further by investigating more efficient design mapping workflow.

References

1. Simonite T (2015) IBM shows off a quantum computing chip. http://www.technologyreview. com/news/537041/ibm-shows-off-a-quantum-computing-chip/
2. Chow JM, Gambetta JM, Magesan E, Abraham DW, Cross AW, Johnson BR, Masluk NA, Ryan CA, Smolin JA, Srinivasan SJ, Steffen M (2014) Implementing a strand of a scalable fault-tolerant quantum computing fabric. Nat Commun 5:4015
3. Kielpinski D, Monroe C, Wineland DJ (2002) Architecture for a large-scale ion-trap quantum computer. Nature 417(6890):709
4. Taylor JM, Petta JR, Johnson AC, Yacoby A, Marcus CM, Lukin MD (2007) Relaxation, dephasing, and quantum control of electron spins in double quantum dots. Phys Rev B 76(3):035315

5. Blais A, Gambetta J, Wallraff A, Schuster DI, Girvin SM, Devoret MH, Schoelkopf RJ (2007) Quantum-information processing with circuit quantum electrodynamics. Phys Rev A 75(3):032329
6. Criger B, Passante G, Park D, Laflamme R (2012) Recent advances in nuclear magnetic resonance quantum information processing. Philos Trans Roy Soc A: Math Phys. Eng Sci 370(1976):4620–4635
7. Grigorenko IA, Khveshchenko DV (2005) Single-step implementation of universal quantum gates. Phys Rev Lett 95(11):110501
8. Saeedi M, Wille R, Drechsler R (2011) Synthesis of quantum circuits for linear nearest neighbor architectures. Quantum Inf Process 10(3):355–377
9. Hirata Y, Nakanishi M, Yamashita S, Nakashima Y (2011) An efficient conversion of quantum circuits to a linear nearest neighbor architecture. Quantum Inf. Comput. 11(1&2):142–166
10. Perkowski M, Lukac M, Shah D, Kameyama M (2011) Synthesis of quantum circuits in linear nearest neighbor model using positive davio lattices. Facta universitatis-series, Electronics and Energetics
11. Chakrabarti A (2011) Sur-Kolay, S., Chaudhury, A.: Linear nearest neighbor synthesis of reversible circuits by graph partitioning. arXiv:1112.0564
12. Wille R, Lye A, Drechsler R (2014) Exact reordering of circuit lines for nearest neighbor quantum architectures. IEEE Trans Comput-Aided Des Integr Circuits Syst 33(12):1818–1831
13. Wille R, Keszocze O, Walter M, Rohrs P, Chattopadhyay A, Drechsler R (2016) Look-ahead schemes for nearest neighbor optimization of 1D and 2D quantum circuits. In: Design automation conference (ASP-DAC), pp 292–297
14. Kole A, Datta K, Sengupta I (2016) A heuristic for linear nearest neighbor realization of quantum circuits by SWAP gate insertion using n-gate lookahead. IEEE J Emerg Sel Topics Circuits Syst 6(1):62–72
15. Zulehner A, Gasser S, Wille R (2017) Exact global reordering for nearest neighbor quantum circuits using A*. In: International conference on reversible computation. Springer, Cham, pp 185–201
16. Shafaei A, Saeedi M, Pedram M (2014) Qubit placement to minimize communication overhead in 2D quantum architectures. In: Design automation conference (ASP-DAC), pp 495–500
17. Lye A, Wille R, Drechsler R (2015) Determining the minimal number of swap gates for multi-dimensional nearest neighbor quantum circuits. In: 2015 20th Asia and South Pacific design automation conference (ASP-DAC), pp 178–183
18. Alfailakawi MG, Ahmad I, Hamdan S (2016) Harmony-search algorithm for 2D nearest neighbor quantum circuits realization. Expert Syst Appl 61:16–27
19. Shrivastwa RR, Datta K, Sengupta I (2015) Fast qubit placement in 2D architecture using nearest neighbor realization. In: 2015 IEEE international symposium on nanoelectronic and information systems (iNIS), pp 95–100
20. Kole A, Datta K, Sengupta I (2018) A new heuristic for n-dimensional nearest neighbor realization of a quantum circuit. IEEE Trans Comput-Aided Des Integr Circuits Syst 37(1):182–192
21. Barenco A, Bennett CH, Cleve R, DiVincenzo DP, Margolus N, Shor P, Sleator T, Smolin JA, Weinfurter H (1995) Elementary gates for quantum computation. Phys Rev A 52(5):3457
22. Miller DM, Wille R, Sasanian Z (2011) Elementary quantum gate realizations for multiple-control Toffoli gates. In: 2011 41st IEEE international symposium on multiple-valued logic (ISMVL), pp 288–293
23. Wille R, Große D, Teuber L, Dueck GW, Drechsler R (2008) RevLib: an online resource for reversible functions and reversible circuits. In: 2008 38th International Symposium on Multiple Valued Logic, ISMVL, pp 220–225

Design of Space-Efficient Nano Router in Reversible Logic with Multilayer Architecture

A. Kamaraj, P. Marichamy, J. Senthil Kumar,
S. Selva Nidhyananthan and C. Kalyana Sundaram

Abstract The reversible logic is a promising technology for zero power dissipation. The reversible logic circuits are realized in quantum cellular automata (QCA), which are called as quantum circuits. The router is a predominant device in the modern communication era. The router is expected to perform faster with minimum area requirement and power consumption. In this paper, a nano router is designed in reversible logic and it is realized in quantum cellular automata. The memory plane and the routing data plane are the major functional components of the router. These are constructed from the basic peripherals decoder, memory array, multiplexer, switch fabric and parallel-to-serial converter. The constructed router peripherals are combined together to form the integrated router and it is realized in QCA. The QCA realized structure has the significance of multilayer crossing for its wire crossing. The multilayer crossing reduces the number of cells required for realization and also it passes the signal without any degradation. The simulation results confirm that the proposed router consumes minimum resources for its realization (up to 50% improvement) than the existing. The nano router is suitable for nano-communication applications. The realization is performed in QCADesigner tool.

A. Kamaraj (✉) · J. Senthil Kumar ·
S. Selva Nidhyananthan · C. Kalyana Sundaram
Mepco Schlenk Engineering College, Sivakasi, India
e-mail: kamarajvlsi@mepcoeng.ac.in

J. Senthil Kumar
e-mail: senthilkumarj@mepcoeng.ac.in

S. Selva Nidhyananthan
e-mail: nidhyan@mepcoeng.ac.in

C. Kalyana Sundaram
e-mail: kalyan@mepcoeng.ac.in

P. Marichamy
PSR Engineering College, Sivakasi, India
e-mail: pmarichamy@psr.edu.in

© Springer Nature Singapore Pte Ltd. 2020
A. K. Singh et al. (eds.), *Design and Testing of Reversible Logic*, Lecture Notes in Electrical Engineering 577, https://doi.org/10.1007/978-981-13-8821-7_13

1   Introduction

Gordon Moore has forecasted, in 1965, that the capacity of an integrated chip grows exponentially with time, whereas the advances in the microelectronic industry depend upon the ever-shrinking size of transistors. The physical limitation of the realization of the transistors is the major challenge in the CMOS technology. The International Technology Roadmap for Semiconductors (ITRS) prediction states that, after 2021 in nanometre technology, it is very hard to shrink the size of the unit transistor [1]. There has been extensive research in recent years at nanoscale to supersede conventional CMOS technology. It is anticipated that these technologies can achieve a density of 1012 devices/cm^2 and operate at THz frequencies [2]. One possible solution is to go with the nanostructure based on quantum-dot cellular automata (QCA).

The significant advantages of the QCA are it has a high packing density, low power-delay product, memory-in-motion and processing by wire (PBW Computation, and Communication occurs simultaneously) [3]. It does not dissipate its signal energy during the transition. Laundauer has proved that the irreversible circuit elements dissipate KTln2 Joules of heat energy (K—Boltzmann constant, T—Room temperature) [4]. Benett has stated that the heat occurs due to the loss of information; which could be avoided by using the reversible logic [5, 6]. The performance factors of the various conventional reversible logic gates are measured and using these gates fundamental combinational circuits are designed [7]. Hence, the low power, high-density computation is possible with the combination of reversible logic circuit realization in QCA.

The router has a data plane and control plane for its function. The control plane has the routing protocols and the data plane does the transfer of data packets. The data plane consists of mux/demux, crossbar and serial-to-parallel converter [8]. There are two types of wire crossings in QCA; named as coplanar and multilayer wire crossing. The coplanar crossing is generally used, but it is suffering from delay, loss of synchronization among signals, discontinuity in propagating signals and more number of cells for its realization. A careful scheduling of clocking is needed to attain the stable output in coplanar crossing. In multilayer, the signal connectivity is more strong and stable and need less control over the clock [9, 10]. Thus, it is desirable to a multilayer crossover in QCA design.

Multistage interconnection networks (MIN) are used in connections in parallel systems to have maximum bandwidth and maximum access rate to memory modules. Here, the switching elements are realized with the multiplexers [11]. The flip-flops are the basic elements of the sequential circuits, which are being constructed with the conventional reversible gates. The flip-flops DFF, SRFF, JKFF and TFF are designed in Master–Slave mode [12]. Two types of memory structures are available: serial and parallel. Serial memory has high latency and multiple-bit storage facility; whereas, parallel memory as low latency and single storage facility. The serial memory is realized in QCA known as SQUARE (Standard Quantum Cellular Automata Array Elements) [13]. As a part of the RAM design, a decoder is designed for accessing the memory. Also, a 2D grid of memory cell is addressed by the QCA decoder [14].

In a bottom-up approach, the Parallel-to-Serial and Serial-to-Parallel converters were designed for data communication purpose. Here, a coplanar based wire crossing is used for communication over the wires [15, 16]. QCA realized sequential circuits are modelled and testing procedure is proposed to reduce the quantum cost and ancilla bits. Simple flip-flops and up–down counter is synthesized and modelled for quantum realization with testing capability [17]. More recently, the reversible circuits design are being promoted towards the universality [18] in 2D rectangular arrays and simulations are done in reversible cellular automata.

The following sections are organized as follows; Sect. 2 describes the required components for the router design, its specifications and its construction in reversible logic. The QCA realization of the proposed architecture is carried out in Sect. 3. A case study and the results discussions are presented in Sect. 4 and is finally concluded.

2   Design Parts of Reversible Router

The basic router has majorly two functional units, they are; a memory block and router block. The memory block consists of a decoder, memory array, multistage interconnection network and multiplexer. The router block has DE multiplexer, switch fabric and a parallel-to-serial converter as shown in Fig. 1. The integral specification of the router is listed in Table 1.

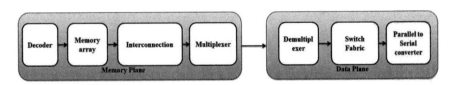

Fig. 1 Functional modules of router

Components of the router	Specification
Table 1 Particularization of the router	
Decoder	2 to 4
Memory array	4 × 4
Multiplexer	4 × 1
DE multiplexer	1 × 4
Parallel-to-serial converter	4 × 1
Switch fabric	4 × 4

Fig. 2 Generic 2 to 4
decoder

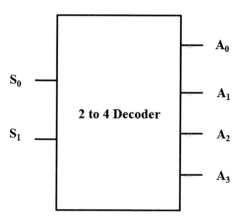

2.1 Decoder

A decoder is a circuit that changes a code (S0, S1) into a set of signals (A0, A1, A2, A3). The basic function is to accept a binary word (code) as an input and create a different binary word as an output. By applying a specified input signal, it is possible to steer the required output. The generic decoder architecture (2 to 4) is shown in Fig. 2 and is switching actions at different time instances are depicted in Fig. 3. It represents A0 is enabled at the time instance 0 and select signals S0 and S1 whose inversion line is closed. Similarly, the other outputs are enabled, when the S0 and S1 are switched at the specified instances. This decoder selects the particular row of memory array for read or write.

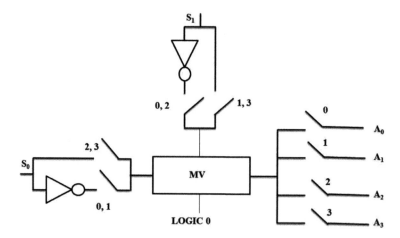

Fig. 3 Switching instances of decoder

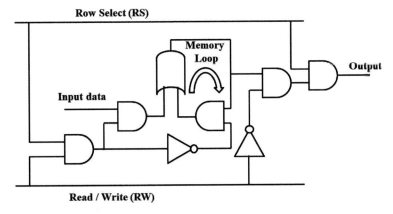

Fig. 4 Single memory cell

2.2 Memory Cell

The arrangement of the single memory cell (1-bit) is depicted in Fig. 4. The row select line enables the particular row of the memory array. The read/write line decide whether the input data to be written (RW = 1) into the memory loop or the memory information to be forwarded to the output (RW = 0). The memory loop retains the value written into the memory cell [14]. Here, the intended design is 4 × 4 array structure; so the single cell has to be replicated 16 times with each row 4 cells.

2.3 Multiplexer and Demultiplexer

A Multiplexer is a circuit with many inputs and one output. By applying a control signal, it is possible to steer any input to the output which satisfies Eq. (1). The switching action instances are shown in Fig. 5, where MV represents the majority voter, A, B, C, D are inputs and S0 and S1 are control signals. The multiplexer selects any one of the memory array data and forward it to the router data plane.

$$Y = (S_0'S_1')A + (S_0'S_1)B + (S_0S_1')C + (S_0S_1)D \tag{1}$$

A Demultiplexer is a circuit with one input and many outputs. The inputs are a data line, two select lines (S0, S1) and a constant enable input (−1) to perform the Demultiplexer operation. The outputs are D0, D1, D2 and D3. By applying a control signal (S0, S1), it is possible to steer any input to the output.

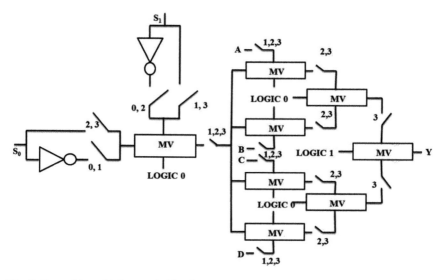

Fig. 5 Reversible multiplexer switching instances

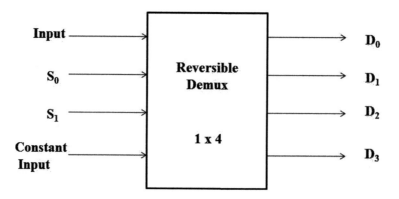

Fig. 6 Reversible demultiplexer

The purpose of 1 × 4 Reversible Demultiplexer is done to perform the process of delivering data packets into the switch fabric as in Fig. 6. One more added advantage is that naturally, the demultiplexer satisfies the reversibility. Demultiplexer forwards the incoming data to the output via selected switch fabric.

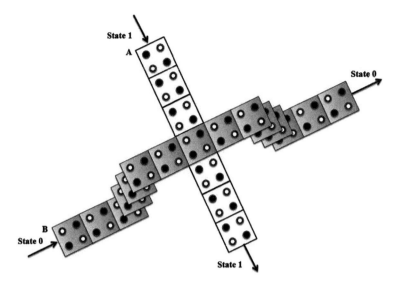

Fig. 7 QCA multilayer crossing

2.4 Switch Fabric

The Switching fabric is to connect input ports to the output ports. The switch fabric includes 4 × 4 interconnections; this is to connect 4-input port (Demultiplexer) and 4-output port (parallel-to-serial converter), thus, switch fabric avoids the collision. This utilizes the multilayer crossing shown in Fig. 7.

2.5 Parallel-to-Serial Converter

In a Parallel-to-Serial Converter, a set of input data are carried by different wires and arrive in the device at the same time. These data are buffered in the horizontal wire by the use of D flip-flop and produced at the output at different instant of time. The switching instances of the parallel-to-serial converter are shown in Fig. 9; where the flip-flops of Fig. 8 are switched at various time instances.

2.6 Integrated Router

The router has two major planes: memory and data planes. The memory plane is constructed using a decoder, memory array and multiplexer. In memory plane, the decoder (2 to 4) acts as a row selector, memory array (4 × 4) is constructed by

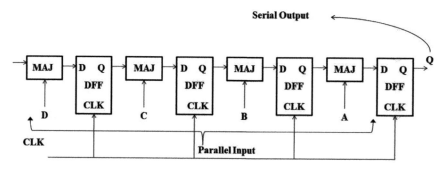

Fig. 8 Parallel-to-serial converter

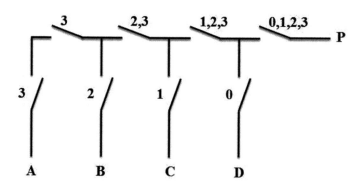

Fig. 9 Switching instances parallel-to-serial Converter

cascading of memory cells in an array and the array is replicated to form the complete memory structure, the multiplexer (4 × 1) functioning as data/bit selector. The row selector selects any one of the rows of data to be forwarded to the output port. The data present in the memory array is available in the data selector through the interconnection network.

The data selector does the forwarding of data serially to the data plane as shown in Fig. 10. The operation of the memory plane is controlled by the sequence of control signals released at distinct time instances.

In data plane, the demultiplexer (1 to 4) receives the data serially from the memory plane. It is sent to the switch fabric for further processing. The switch fabric receives data in parallel from all the demultiplexers. This parallel data is forwarded to the output port serially via parallel-to-serial converter. The switching actions at various time instances control the entire operation to be performed without any confusion in data forwarding. The component level detailed representation is shown in Fig. 11, which is being controlled by the switching actions as shown in Figs. 3, 5 and 9.

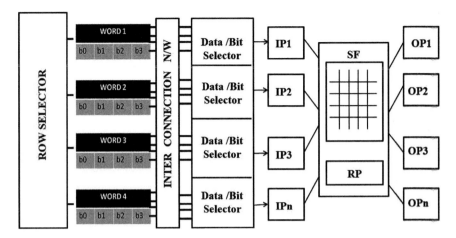

Fig. 10 Functional representation of router

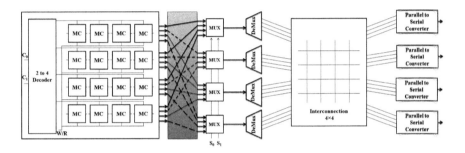

Fig. 11 Components level description of router

3   QCA Realization and Functional Verification

In Sect. 1, it has been discussed that the multilayer architecture provides better data communication than its counterpart coplanar. The reversible gates Feynman, Fredkin, Toffoli and Peres gates are realized in QCA using majority voter gate with multilayer (crossover and vertical cell used—bridge type) [19] and coplanar (45° rotation in horizontal wire for crossing) architecture. The realization and simulation results confirm that the multilayer architecture is having a fewer number of cells, area, and simulation runtime than coplanar as shown in Table 2. Also, the coplanar structure requires careful alignment of cells and clocking to have proper propagation of signals.

Table 2 Comparison of coplanar and multilayer architecture

Parameters	Feynman gate		Fredkin gate		Toffoli gate		Peres gate	
	Coplanar	Multilayer	Coplanar	Multilayer	Coplanar	Multilayer	Coplanar	Multilayer
No. of cells	122	77	392	133	131	90	310	194
Area-nm^2	39,528	24,948	127,008	43,092	42,444	29,160	100,440	62,856
Simulation time (s)	34	27	96	45	45	35	87	65

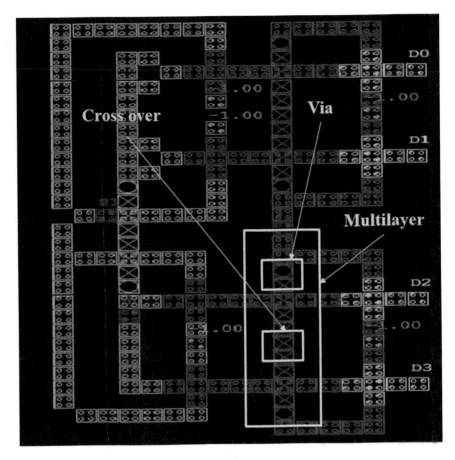

Fig. 12 Demultiplexer

The major components of the router are realized in the QCA environment and their functionality is being verified. Here, the crossover is structured in multilayer configuration and the same is indicated in Figs. 12, 13, 14, 15, 16 and 17. Also, as discussed in Sect. 2, the router has memory and data plane; whose realization is shown in Fig. 16, 17 and the complete integrated router architecture is shown in Fig. 18. The integral part of the designed router is also indicated.

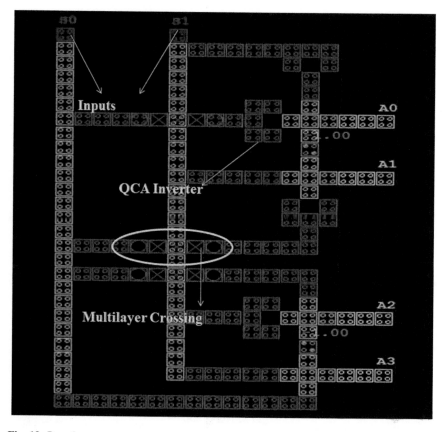

Fig. 13 Decoder

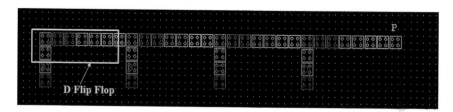

Fig. 14 Parallel-to-serial converter

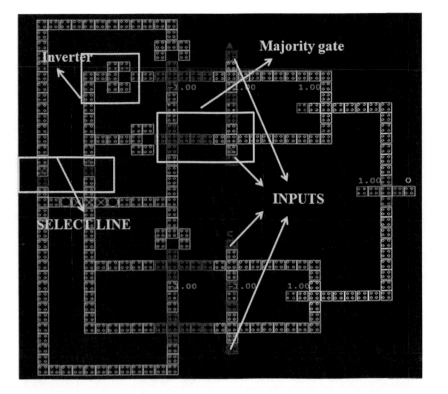

Fig. 15 Multiplexer

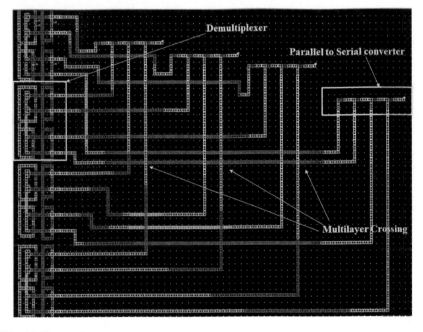

Fig. 16 Router plane structure

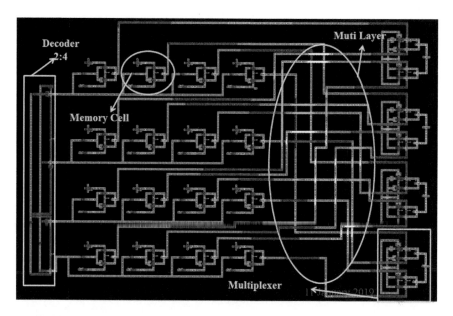

Fig. 17 Memory plane structure

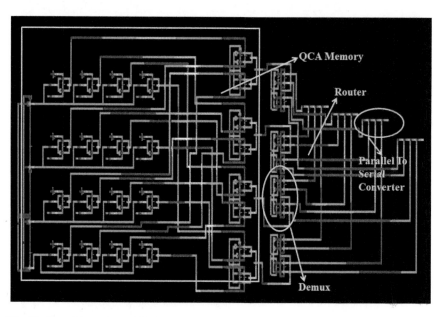

Fig. 18 Complete router architecture

4 Results and Discussion

Ratification of the Router was done in different aspects and are revealed as below.

Case 1
The decoder is provided with select lines as '00' and Read/Write line as '1111' which performs its respective operations as shown in the simulation output Fig. 19.

Case 2
The decoder is provided with select lines as '10' and Read/Write line as '1010' which performs its respective operations as shown in the simulation output Fig. 20.

The router is realized in multilayer crossing, its individual components resource utilization is listed in Table 3. It is observed that there are three layers are required for realizing the router architecture. The coplanar realization requires 45° rotation of horizontal cells wherever the crossing is required, and also it requires sufficient clocking sequence for polarization of the cells without interference [20, 21]. In multilayer crossing, it requires only crossover cells and vertical cells to make the crossing.

The proposed router design is compared with the existing router [8] design. In the proposed design, the minimum amount of resources is utilized for the router realization due to the proper clock scheduling and organization. Here, no needs of additional cells for signal boosting as in the case of coplanar. Thus, the improvement in complexity and area is achieved for the integrated router design in QCA as shown in Table 4.

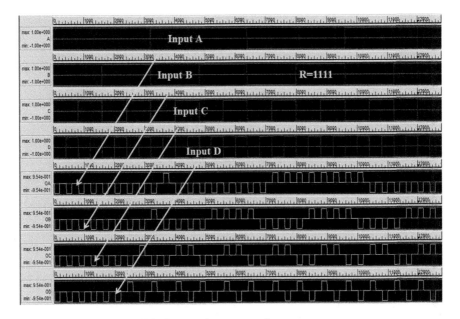

Fig. 19 Simulation result of the integrated router case 1

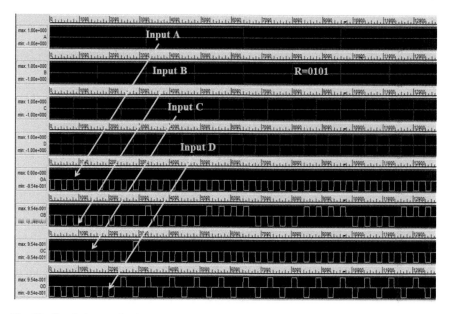

Fig. 20 Simulation result of the integrated router case 2

Table 3 Resource utilization of router components

Parameters	Demux	P to S converter	Router	Decoder	Mux	Memory	Complete router
Complexity	22 cells	41 cells	3057 cells	188 cells	277 cells	7465 cells	10,932 cells
Simulation time (s)	4	4	25	4	5	75	128
Area (μm²)	212,400	50,150	7,910,900	203,500	448,500	16,747,500	30,394,500
Latency (s)	1	4	6	3	5	20	44
Layers	3	1	3	3	3	3	3

Table 4 Comparison of existing and the proposed router

Parameters	Router existing design [8]	Router proposed design	Percentage of improvement
Complexity	4026	3057 cells	24%
Latency	48	24	50%
Area (μm²)	13.81	7.91	42.7%
Simulation time	–	25 s	–
Layers	3	3	–

5   Conclusion

The design of nano router architecture in reversible logic and realizing them in quantum cellular automata is one of the viable solutions in order to achieve the low-power dissipation and high-speed device. Here, an integrated router with memory plane and control plane is designed using the combination of digital components decoder, memory cell, multiplexer, demultiplexer and parallel-to-serial converter. The designed nano router structure is realized in the QCA platform with multilayer architecture for wire crossing. The multilayer crossing effectively propagates and computes the information while propagation without any interferences in the signal. The simulation work is carried on in QCADesigner 2.0.3. From the simulation and realization results, it is evident that the reversible logic based router realization in QCA with multilayer architecture effectively improves the performance up to 50% than the existing method. The designed router could be deployed in communication and networking environment in the near future.

References

1. Jeong K, Kahng AB (2009) A power-constrained MPU roadmap for the international technology roadmap for semiconductors (ITRS). IEEE https://doi.org/10.1109/socdc.2009.5423856
2. Tahoori MB, Lombardi F (2004) Testing of quantum dot cellular automata based designs. In: Design, automation and test in Europe conference and exhibition, vol 2, pp 1408–1409
3. Amlani I et al (1998) Demonstration of a six-dot quantum cellular automata system. Appl Phys Lett 72(17):2179–2181
4. Landauer R (1961) Irreversibility and heat generation in the computational process. IBM J Res Dev 5:183–191
5. Bennett CH (1973) Logical reversibility of computation. IBM J Res Dev 17:525–532
6. Fredkin E, Toffoli T (1982) Conservative logic. Int J Theory Phys 21:219–253
7. Anamika, Bhardwaj R (2018) Reversible logic gates and its performances. In: Proceedings of the second international conference on inventive systems and control (ICISC 2018). IEEE Explore. https://doi.org/10.1109/icisc.2018.8399068
8. Sardhinha LHB, Costa AMM, Neto OPV, Vieira LFM, Vieira MAM (2013) Nano router: a quantum-dot cellular automata design. IEEE J Sel Areas Commun 31(12): 825–834. Supplement-part 2
9. Lu L, Liu W, O'Neill M, Swartzlander EE (2013) QCA systolic array design. IEEE Trans. Comput. 62(3):548–560
10. Graunke CR, Wheeler DI, Tougaw D, Will JD (2005) Implementation of crossbar network using quantum-dot cellular automata. IEEE Trans Nanotechnol 44:435–440
11. Tehrani MA, Safaei F, Moaiyeri MH, Navi K (2011) Design and implementation of multistage interconnection networks using quantum-dot cellular automata. Microelectron J 42:913–922
12. Singh PL, Majumder A, Chowdhury B, Mondal AJ, Shekhawat TS (2015) Reducing delay and quantum cost in the novel design of reversible memory elements. Procedia Comput Sci 57:18–198

13. Ottavi M, Vankamamidi V, Lombardi F, Pontarelli S (2005) Novel memory designs for QCA implementation. In: Proceedings of 2005 5th IEEE conference on nanotechnology, vol 5, pp 1–4
14. Walus K, Vetteth A, Jullien GA, Dimitrov VS (2003) RAM design using quantum-dot cellular automata. In: Technical proceedings of nano technology conference and trade show, vol 2, pp 1–4
15. Govindapriya K, Periyasamy M (2016) Nano design of communication parts with QCA using reversible logic. In: International conference on advanced communication control and computing technologies, vol 13, pp 20–33
16. Kannaki S, Karthigai Lakshmi S, Kaviya G (2018) Design and performance analysis of serial communication in QCA. Int. J. Pure Appl. Math. 119(12):825–836
17. Khan MHA, Jacqueline E (2018) Rice: first steps in creating online testable reversible sequential circuits. VLSI Design 10: 1–13. Hindawi
18. Adamatzky A (2018) Reversibility and universality. Springer, Berlin
19. Biswas P, Gupta N, Patidar N (2014) Basic reversible logic gates and its QCA implementation. Int J Eng Res Appl 4(6):12–16
20. Tougaw PD, Lent L (1994) Logical devices implemented using quantum cellular automata. J. Appl. Phys. 75(3):1818–1825
21. Shin S-H, Jeon J-C (2014) Kee-young yoo: design of wire crossing technique based on difference of cell state in quantum-dot cellular automata. Int J Control Autom 7(4):153–164

Design of Reversible Binary-to-Gray Code Converter in Quantum-Dot Cellular Automata

I. Gassoumi, L. Touil and B. Ouni

Abstract At nanoscale, for digital systems, the device density and power constraint of the circuit are essential issues. Quantum-dot cellular automata (QCA) is an incipient nanotechnology, which leads to build circuits at nanoscale. It offers various features such as minimal power dissipation, very high-operating frequency, and nanoscale feature size. Besides, reversible computation can lead to the development of low-power systems without loss of information. Thus, reversible QCA logic can provide a powerful and efficient computing platform for digital applications. This paper presents a QCA code converter. Feynman gate is used as a fundamental building block to perform the proposed design of code converter. QCADesigner version 2.0.3 is used to validate the accuracy of the proposed circuit. QCAPro, a very widespread power estimator simulation engine, is applied to estimate the power depletion of the proposed circuit.

1 Introduction

The thermal energy discharged by circuit transistors is one of the most important issues faced while designing VLSI circuits. Although the recent VLSI technology is based on CMOS technology. However, this technology entails many challenges. High-power density levels, high leakage currents, and constraint of speed in GHz are some of the problems that CMOS is faced [1, 2]. In the 1960s, Landauer proved

I. Gassoumi (✉) · L. Touil
Laboratory of Electronics and Microelectronics,
University of Monastir, Monastir, Tunisia
e-mail: gassoumiismail@gmail.com

L. Touil
e-mail: lamjedtl@yahoo.fr

B. Ouni
Higher Institute of Technologies of Sousse,
University of Sousse, Sousse, Tunisia
e-mail: ouni_bouraoui@yahoo.fr

© Springer Nature Singapore Pte Ltd. 2020
A. K. Singh et al. (eds.), *Design and Testing of Reversible Logic*, Lecture Notes in Electrical Engineering 577, https://doi.org/10.1007/978-981-13-8821-7_14

that if information processing has performed at an irreversible process, then energy will be lost and a loss of information leads to loss in energy [3]. It has proved that the loss of any bit of information would be at least 0.6931 KbT joules, where T is the temperature of the environment in which computation is performed. These truths lead to look beyond traditional approaches in order to reduce the power depletion. Hence, various alternative technologies were found like Carbon Nanotube Field-effect Transistors (CNTFET), Single-electron Transistor (SET), and Resonant-tunneling Diodes (RTD) [4].

QCA is an incipient nanotechnology and potential substitute to orthodox CMOS archetypes at nano extent that assures to conceive digital circuits with minimal power, extreme speed, and particularly dense structures. This approach can operate at a higher frequency (in the order of THz) than the conventional solution [5–7]. Apart from that, reversible approach is also considered as an alternative technology that can mitigate the issues, which are anticipated for CMOS devices due to the heat dissipation [8, 9]. A system is reversible if its output values define its input values, i.e., it performs a bijective function. The inputs may be reproduced from the outputs and vice versa. Thus, without wasting of information, no energy dissipation occurs. This design strategy aims toward the development of digital designs with ideally zero power dissipation. Then, the design of reversible logic circuits used in this new emerging nanotechnology (QCA) leads to ultra-low-power systems and architectures. Reversible QCA logic can provide high-performance and low-power solutions for digital designs. Recently, it is used as the best method to reduce power depletion [10–15]. As well known that the fundamental components of each logic circuits are logic gates which are used to perform any Boolean functions. Thus, reversible circuits can easily be realized that can perform complex logical and arithmetic operations using reversible logic gates. Some gates have been employed by several reversible logic designs such as Feynman, Fredkin, and Toffoli gates. In recent years, several efforts have been made toward to conceive of reversible QCA digital circuits [16–18].

This chapter describes the design of reversible binary-to-gray code converter in QCA. The basic building block of the proposed design is Feynman gate. In this chapter, the major contributions of our work can be summarized as follows:

- A novel reversible Feynman gate (FG) based on QCA technology has been proposed.
- The proposed FG has been exploited to realize the design of reversible code converter circuit in QCA technology.
- The simulation results of the designed circuit have been correctly obtained.
- Power dissipation of the proposed design has been estimated.

The rest of this work is systematized as follows: Sect. 2 reviews the QCA technology. Section 3 presents the background work. The designed circuit is discussed in Sect. 4. Section 5 shows the performance comparisons with power depletion analysis. Finally, the chapter is concluded in Sect. 6.

2 Background

2.1 QCA Devices

The QCA approach is an emerging technology designed as an appropriate alternative to conventional technology. It contains an array of cells. Cells in the QCA technology include four cavities located in four corners of the square. Only two electrons diametrically opposite are injected into a cell due to Coulombic interaction [19]. Through Coulombic effects, two possible polarizations (labeled –1 and 1) can be shaped. These polarizations are represented by binary "0" and binary "1" as depicted in Fig. 1. Figure 2 illustrates the propagation of logic "0" and logic "1", respectively, between input and output in QCA binary wires due to the Coulombic repulsion. Generally, in neighboring cells, the Coulombic interaction between electrons is used to implement many logic functions, which are controlled by the clocking mechanism [20].

The main logic components in the QCA technology are majority and inverter gates, which are composed by some QCA cells as depicted in Fig. 3 [21, 22]. The 3-input majority voter consists of five QCA cells. Furthermore, the majority gate and inverter are operated for realizing QCA circuit performances. A majority gate can operate a 3- or 5-input logic functions as presented in Fig. 3.

Fig. 1 Two different polarization of quantum-dot cell

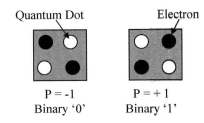

Fig. 2 QCA binary wire

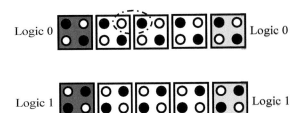

(a)   **(b)**

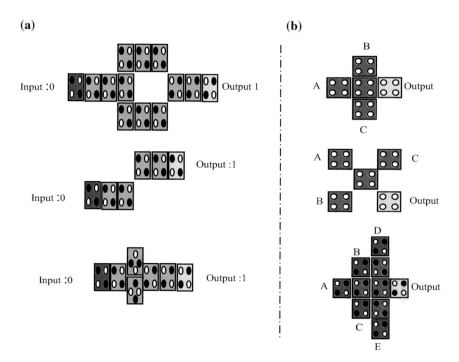

Fig. 3 **a** Types of inverter gate, **b** majority gate

2.2 QCA Clocking

Clocking plays an important role in QCA designs. Every single clock has four periods, namely, switch, hold, release, and relax, which are essential for appropriate circuit implementations as illustrated in Fig. 4 [23, 24].

- In switch phase, QCA cell is beginning to shift from unpolarized status to polarize status, and the blockades of the dots are lifted.
- In hold phase, blockade of the cell is in the highest value, electron cannot channel within dots, and cell preserves their existing statuses specifically, stable polarization.
- In release phase, the blockade is lessened, electron can channel within dots, and statuses of the cell turn into unpolarized.
- In the final phase, blockade stays lessened and cell remains in unpolarized status.

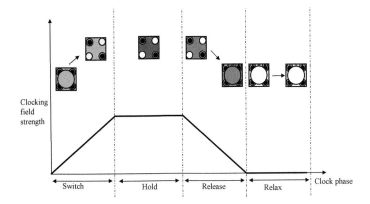

Fig. 4 QCA clock zones

3 Feynman Gate

The reversible designs are one of the most promising solutions, which are capable of overcoming the limitations of the applications based on the CMOS technology. Additionally, reductions in energy dissipation comprise one of the important goals of nanotechnology-based methods, including QCA and so it is desirable to consider reversibility in the design of QCA circuits. The Feynman gate (FG) is employed by many reversible logic circuits. It is a 2×2 universal gate (Fig. 5) and any logical reversible design can easily be implemented by using it. The FG is named as a self-inverse one, i.e., to a circuit that is capable to return to any previous state in reverse order. The truth table of the Feynman gate is shown in Table 1. Figure 6 illustrates the proposed QCA design and the simulated outcome of Feynman gate.

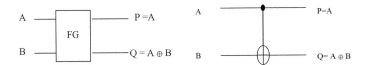

Fig. 5 Feynman gate and its quantum realization

Table 1 Truth table of FG

A	B	P	Q
0	0	0	0
0	1	0	1
1	0	1	1
1	1	1	0

(a)   **(b)**

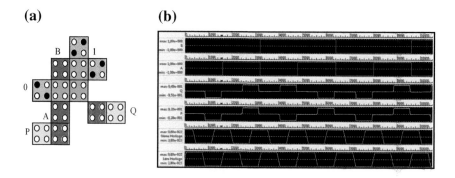

Fig. 6 **a** Proposed QCA layout of Feynman gate, **b** its simulation outcome

The proposed Feynman gate contains 12 QCA cells and covers an area of 0.01 μm^2. Here, the suppression of majority voter leads to QCA structures with less power consumption and hardware complexity.

4   Reversible Binary-to-Gray Code Converter

In this section, a new reversible code converter design by using QCA is explored. A binary-to-gray Code converter is a combinational circuit which is a non-weighted code. This conversion method is useful to reduce the rapid switching activity. In order to generate the 3-bit binary-to-gray code converter, only two FGs have been used. The corresponding block representation and QCA diagram of the proposed circuit is depicted in Figs. 7 and 8, respectively. It has one garbage bit. The proposed design contains 29 cells, extent 0.04 μm^2. It needs two clock zones to generate the correct outputs.

Fig. 7 Schematics of 3-bit reversible binary-to-gray code converter

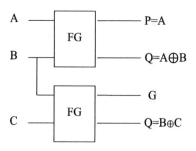

(a)

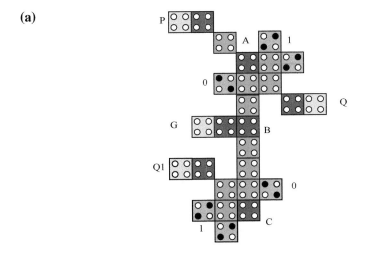

(b)

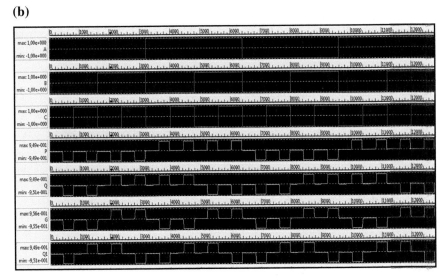

Fig. 8 a QCA layout of 3-bit reversible binary-to-gray code converter, **b** its Simulation outcome

5 Results and Discussions

QCADesigner software is used to verify and simulate the proposed hardware design [25]. The utilized parameters for the simulation are shown in Table 2. The Feynman gate is the basic building blocks of the proposed circuit. Therefore, the foremost benefit of the designed gate is that no multilayer and rotating crossing is applied, which will lead to the efficient design of reversible binary-to-gray code converter. Table 2 illustrates the comparison result of the designed gate (FG) with some

Table 2 Simulation parameters

Parameter	Value
Number of samples	12,800
Convergence tolerance	0.001000
Radius of effect	65,000,000 (nm)
Relative permittivity	12,900,000
Clock low	3,800,000e-023(J)
Clock high	9,800,000e-022(J)
Clock shift	0,000,000e + 000
Clock amplitude factor	2,000,000
Layer separation	11,500,000
Maximum iterations per sample	100

existing designs in the literature. It can be perceived that the proposed gate excels all the best reported designs presented in [26]. Table 3 depicts the quantum cost (QC) of the suggested circuits. Clearly, the QCA designs have less QC than that of classical implementation as shown in Table 4. In addition, the proposed design consists of 29 cells, occupies 0.04 μm^2 area and 0.5 latency. The QCAPro software, a probabilistic designing engine [27], has been applied for power depletion study.

Table 3 Comparison of various Feynman gates

Design	Cell count	Area (μm^2)	Clock no. cycle
Feynman gate [11]	54	0.037	0.75
Feynman gate [14]	43	0.038	0.75
Feynman gate [15]	13	0.02	0.5
Feynman gate [26]	14	0.01	0.5
Proposed Feynman gate	12	0.01	0.5
3-bit reversible binary-to-gray code converter [28]	118	0.0926	4
3-bit reversible binary-to-gray code converter [29]	117	0.0953	10
3-bit reversible binary-to-gray code converter [30]	75	0.0554	4
Proposed binary-to-gray code converter	29	0.04	0.5

Table 4 Quantum cost of the proposed reversible sub-circuit versus corresponding QCA layout

Proposed reversible circuit	Quantum cost	Quantum cost of QCA circuit (area. latency^2)
Feynman gate	1	0.0025
Reversible binary-to-gray code converter	2	0.01

Table 5 Power dissipation analysis of the proposed circuits

Proposed QCA circuit	Dissipation of power T = 2.0 K		
	$\gamma = 0.5$ Ek	$\gamma = 1$ Ek	$\gamma = 1.5$ Ek
Feynman gate	9.55	15.76	22.73
Reversible binary-to-gray code converter	19.39	31.52	45.98

Fig. 9 The effect of temperature on output polarization of Feynman gate

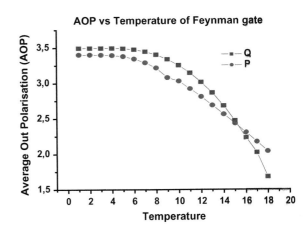

Table 5 explains the overall power depletion study of the proposed design. The estimation is performed at three tunneling energy levels (EK) at T = 2 K. The temperature impact on the output polarization of designed gate is performed. The proposed diagram function readily in the temperature range of 1–6 K, and the AOP for each cell is reformed very little in this estimate as demonstrated in Fig. 9, which is analyzed at different temperatures by QCADesigner Tool.

6 Conclusion

Recently, QCA technology has attracted researchers' attention for implementing reversible computing. Reversible QCA logic has become a promising technology in the implementation of digital design. In this chapter, we have presented a QCA architecture based on reversible logic, which performs binary-to-gray code converter. Feynman gate is used to achieve the proposed design. The results of the comparison demonstrated significant improvements. The designed circuit is more efficient in terms of extent, cell complexity, quantum cost, and delay.

References

1. Haron NZ, Hamdioui S (2008) Why is CMOS scaling coming to an END? In: 3rd IEEE international design and test workshop (IDT), pp 98–103
2. Bondy PK (2002) Moore's law governs the silicon revolution. In: Proceedings of the IEEE, pp 78–81
3. Landauer R (1961) Irreversibility and heat generation in computing process. IBM J Res Dev 5 (3):183–191
4. ITRS, International Technology Roadmap for Semiconductors 2013 Update Technical Report. Available www.itrs.net
5. Niamat M, Panuganti S, Raviraj T (2010) QCA design and implementation of SRAM based FPGA configurable logic block. In: 53rd IEEE international midwest symposium on circuits and systems (MWSCAS), pp 837–840
6. Zhang R, Walus K, Wang W, Jullien GA (2004) A method of majority logic reduction for quantum cellular automata. IEEE Trans Nanotechnol 3(4):443–450
7. Orlov AO, Amlani I, Bernstein GH, Lent CS, Snider GL (1997) Realization of a functional cell for quantum-dot cellular automata. Science 277(5328):928–930
8. Bennett CH (1973) Logical reversibility of computation. IBM J Res Dev 17(6):525–532
9. Wille R, Soeken M, Miller M, Drechsler R (2014) Trading off circuit lines and gate costs in the synthesis of reversible logic. Integr VLSI J 47(2):284–294
10. Sasamal TN, Mohan A, Singh AK (2018) Efficient design of reversible logic ALU using coplanar quantum-dot cellular automata. J Circuits Syst Comput 1–19
11. Das JC, De D (2016) Novel low power reversible binary incrementer design using quantum-dot cellular automata. Microprocess Microsyst 42:10–23
12. Debnath B (2016) Reversible logic-based image steganography using quantum dot cellular automata for secure nanocommunication. IET Circuits Devices Syst 1–10
13. Das JC, De D (2016) User authentication based on quantum-dot cellular automata using reversible logic for secure nanocommunication. Arab J Sci Eng 41(3):773–784
14. Das JC, De D (2016) Quantum dot-cellular automata based reversible low power parity generator and parity checker design for nanocommunication. Front Inf Technol Electron Eng 17(3) 224–236
15. Bhoi (2017) Design and evaluation of an efficient parity-preserving reversible QCA gate with online testability. Cogent Eng 1–18
16. Ahmad PZ (2017) A novel reversible logic gate and its systematic approach to implement cost-efficient arithmetic logic circuits using QCA. Data Brief 15:701–708
17. Chabi AM (2017) Towards ultra-efficient QCA reversible circuits. Microprocess Microsyst 49:127–138
18. Kamaraj A, Ramya S (2014) Design of router using reversible logic in quantum cellular automata. In: International conference on communication and network technologies (ICCNT), pp 249–253
19. Lent CS (1993) Quantum cellular automata. Nanotechnology 4(1):49–57
20. Tougaw PD, Lent CS (1994) Logical devices implemented using quantum cellular automata. J Appl Phys 75(3):1818–1825
21. Lent CS (2003) Clocked molecular quantum-dot cellular automata. IEEE Trans Electron Devices
22. Tang R, Zhang F, Kim YB (2005) Quantum-dot cellular automata SPICE macro model. In: Proceedings of the 15th ACM great lakes symposium on VLSI, pp 108–111
23. Momenzadeh M, Huang J, Lombardi F (2008) Design and test of digital circuits by quantum-dot cellular automata. Artech House, pp 37–67
24. Angizi S, Sarmadi S, Sayedsalehi S, Navi K (2015) Design and evaluation of new majority gate-based RAM cell in quantum-dot cellular automata. Microelectron J 46(1):43–51
25. Walus K, Dysart TJ, Jullien GA, Budiman RA (2004) QCADesigner: a rapid design and simulation tool for quantum-dot cellular automata. IEEE Trans. Nanotechnol 3(1):26–31

26. Biswas PK, Bahar AN, Habib MA, Al-Shafi MA (2017) Efficient design of Feynman and Toffoli gate in quantum dot cellular automata (QCA) with energy dissipation analysis. Nanosci Nanotechnol 7(2):27–33
27. Srivastava S (2011) QCAPro – an error-power estimation tool for QCA circuit design. In: International symposium of circuits and systems (ISCAS), pp 2377–2380
28. Das JC, De D (2015) IETE J. Res. 1
29. Neeraj KM, Subodh W, Singh VK (2016) In: Proceedings of the 4th international conference on frontiers in intelligent computing: theory and applications (FICTA)
30. Javeed IR, Banday MT (2011) Efficient design of reversible code converters using quantum dot cellular automata. J Nano- Electron Phys 1–8

Index

© Springer Nature Singapore Pte Ltd. 2020
A. K. Singh et al. (eds.), *Design and Testing of Reversible Logic*, Lecture Notes
in Electrical Engineering 577, https://doi.org/10.1007/978-981-13-8821-7